To my dear parents

Also by the author in English

Aura and Consciousness - New Stage of Scientific Understanding. 1998.

Light after Life. 1998

Human Energy Fields: Study with GDV Bioelectrography. 2002.

Measuring Energy Fields: State of the Art. GDV Bioelectrography series. Vol. I. Korotkov K. (Ed.). 2004.

You may order these books from
www.backbonepublishing.com

Konstantin Korotkov

SPIRAL TRAVERSE
Journey into the Unknown

SPIRAL TRAVERSE
Journey into the Unknown

Konstantin Korotkov

This book describes spiritual adventures with mediums, healers, and ordinary people on various continents. It is written by a scientist of world-renown, who has devoted his life to the study of spiritual worlds from a scientific perspective. Science, Information, and Spirit - this is a recurring slogan of his work for many years, and also is the name for annual international congresses held in Saint-Petersburg every July. This is a book about unusual situations in the world around us, and about unusual lives of apparently ordinary people in Russia, the USA, Colombia, and Venezuela. Expeditions to different parts of the world to measure the energy of Space, measurements of health, prayer and love - these are only some topics of these studies. Life - is a great adventure in our everyday reality, and you may receive a strong impulse of optimism by reading this book.

Translated from Russian by the author and Veronica Kirillova
Edited by Berney Williams and Ludmila Tolubeeva

ISBN-13: 978-0-9742019-4-8
ISBN-10: 0-9742019-4-4

Library of Congress Control Number: 2006926838

Backbone Publishing Company
P.O. Box 562, Fair Lawn, NJ 07410, USA
FAX: 201 670 7892
bbpub@optonline.net
www.backbonepublishing.com

CONTENTS

Traverse - the type of mountaineering experience where athletes ascend one peak after another without lowering down to the ground level.

EPC bioelectrography - Electrophotonic Camera, based on Gas Discharge Visualization (GDV) technique.

Part I

ENIGMAS OF LIFE LIGHT

INTRODUCTION

I have friends and colleagues: Bob Van de Castle and Justine Owens. Both are Americans, professors at the University of Virginia, founded by Thomas Jefferson - one of the fathers of the American constitution. Bob is an outstanding world specialist in dream research; Justine has studied psychophysiology all her life. They are scientists well-known in the world, and American in understanding.

In the US one can often meet friendly and sympathetic people, sincerely ready to help, deeply sensitive and emotional. This is the effect of religious education which has been the basis of social life for two centuries.

It was very interesting to meet, talk, and travel together with Bob and Justine. And since both of them are interested and professionally work with frontier phenomena of the human psyche, something unusual is always going on around both of them. The circumstances under which they first met were most unusual.

When Justine appeared at the University of Virginia as a young physiologist, Bob had already been a venerable professor. They met each other a couple of times at official gatherings, but no serious acquaintance. And one night Justine had a dream.

Phone is calling. She takes the phone and hears,

"Hello, Justine. This is professor Van de Castle. I am starting a new project on dream research. Would you like to take part as a physiologist? We will be investigating very interesting states of human consciousness; this work will continue for many years and will change your life totally. Then we will start using a new, yet non-existing device for the study of human energy field."

Justine replied something (in her sleep), but when she woke up she did not recall any other details. She was lying in bed, thinking of the meaning of such a dream: they met with professor Van de Castle only a couple of times and there was no question about joint work. Moreover, she had enough of her own work to do. But this moment the telephone rang. Justine took the receiver and heard,

"Hello, Justine. This is professor Van de Castle. I am starting a new project on dream research. Would you like to take part as a physiologist? We will be investigating very interesting states of human consciousness, and I hope this work will continue for many years."

From this moment on they began their long work, which really changed Justine's fate. Bob is a very interesting person, wonderful story-teller, a visionary, super active, and being in his mid-seventies is energetically surpassing most of the middle-age people. Therefore, I was pleased to agree to give a talk at the University of Virginia and travel around to interesting places of the state.

And as to the device from Justine's dream, we will touch upon it later.

One day we went to visit the Temple of Light. The way curved in the hills, very similar to our Russian region Kostroma. The main difference was in the quality of

Fig.1. The Temple of Light.

Fig.2. The Hall of Light.

Fig.3. Wooden plate with carved citations.

Fig.4. The Hall of Light.

asphalt: Americans somehow manage to keep it without holes and ruts. What is more, constantly, not only before the visit of the country governor. Sparsely populated areas of the state were striking. Three hours by car from Washington D.C. and you can drive in the forests and hills several hours and meet no built-up areas. Only cars slowly going in different directions. As a matter of fact, with our demographical situation in Russia we will have the same thing soon all over the country.

We strayed fairly long in the hills, and, finally, came to some buildings.

"Looks as if we've arrived", said Bob and parked the car.

"Are you sure that it is here?" asked Justine, "There is no parking place."

There was a lot of free space around, but in the USA any public place is first of all equipped with good parking, and its size determines the rank of the place. Just like our public places in Russia can be evaluated by the quality of restrooms.

We got out of the car, and, having looked around, Bob guided us to some large veranda. When we came closer we realized that this was not a simple veranda. A human-size statue of the dancing Indian god Shiva was standing in the middle in a huge glass case, and statue of an old man with a long beard sitting in Oriental clothing was rising on the other side.

"This is the founder of the Yoga Center, where we are now", explained Bob.

"Is this his tomb?" I asked fondly.

"No, he is still in good health, this is his copy in his lifetime", replied Justine. I wanted to knock at the statue to understand if it was made of metal or plastic, but hesitated. Americans have a big knowledge in fiberglass Disney lands.

Passing along the veranda we found ourselves on the edge of the hill from where a view to the Temple of Light opened up before our eyes. The place was wonderfully chosen. The temple was standing on the bank of a small lake, in a dell, surrounded by hills. It looked like a ship sailing in the blue waters from the top of our hill.

"As I suspected we got onto a wrong road", said Bob, pointing at the narrow motor way, coming to the gates of the temple complex.

"But if we look for a bypass route, another hour will be lost", said Justine, "So, I suggest going down here."

Exactly what we did, running down the steep grassy slope.

The building of the Temple was under restoration and covered with canvas. According to the design it should have represented a lotus. We entered the temple and got into a big hall. A sculpture with a shining sphere, obviously symbolizing the world globe, was standing in the center and 18 glass cases were located around the hall. Each was dedicated to some world religion, and sacred books, statues and images most characteristic of these religions were displayed there (Fig. 1, 2)

At the second flour we found another big hall. Wooden plates with carved dicta from sacred texts of various religions were placed in the stands around the walls. Each dictum was dedicated to the light, as a personification of the essence of God. A ray of light was rising in the middle of the hall from the pedestal surrounded by lotuses and dispersed under the ceiling to the symbols of each religion. This was the embodiment of the one God, feeding all nations with his light, giving a ray of Truth to the Earth, refracted in different ways for different nations, but keeping its luminiferous vibration. I was not lazy to take pictures of all these plates and used the dicta as epigraphs to the next chapters of this book. The author disavows himself to be incorrect: I quote what was written (Fig. 3,4).

Why all nations have always associated God with light? Is it an accidental fact that the heads of saints in Christian icons are surrounded by nimbuses - the glow which has always been discussed by righteous people, observing sacred phenomena? Pictures of such glow can be found in the images of Indian and South American Gods. Is this a fantasy, the Aura, or the notorious biofield, laid bare by our academic committees? What do extrasensory individuals see then? Does the biofield exist? Or it is "opiate of the people", the tales of cunning charlatans rising time and again within thousands of years? Let us try to leisurely investigate these, as well as many other questions.

SCIENCE, INFORMATION, SPIRIT

"And waters will burst the banks, and will drown blooming cities and fertile lands. Marble floors will be covered with river mud, the fish will be swashing in the streets and squares of the capitals, and sun beaches will be covered with snow. Catastrophes will be following one another in the world, the peak of cataclysms will fall on year 2003-2004, after which will slow down".

Or, in other words: "The nearest interference of influence of the given aggregate of cosmo-planet phenomena upon the Earth is expected in 1999 - 2006 in the form of exceptionally negative effects for the subjects of ethno and biosphere" 1.

Such prophecies were made public by Russian geologist professor V. Rudnik in 2000 at the IV International Congress "Science. Information. Spirit" based on the analysis of astrological maps. Then everybody listened, discussed, and forgot in a week, but with every coming year the scale of catastrophes seemed to be more and more like it had been predicted. Well, it totally coincided in the years 2002-2005. The catastrophic flooding in Europe, USA and Asia, cascades of waters, whirling in the streets, washed away and toppled over cars, hundreds of the dead and missing.

Typhoons, hurricanes, earthquakes all over the world. Plus human madness culminated in terrorism and wars. Bio-sphere became wild just as human brains. Different aspects of our Earth environment under the influence of the Universe activity.

Of course, it is difficult to realize the scale of such catastrophes sitting in a comfortable apartment in St. Petersburg, Moscow, or Washington. What is more, sitting in a decorous conference hall. Besides, the futurologists always attach globally universal character to their predictions. We are lucky that they have local character.

But this was only one of many topics discussed at the congress "Science. Information. Spirit." within the last 10 years (Fig. 5)

Megalithic structures of Europe and America, chronal and argon generators, psychology of communication with the unborn and dead, therapy with mineral, gemstones and leaches, express-diagnostics of a wide range of diseases. What is the connecting point between these topics, seeming cardinally opposite? Studying the widest range of traditional, classic, and absolutely fantastic directions authors from more than 40 countries of the world use one and the same method - the method of **EPC bioelectrography, Evoked Photon Capture** based on Gas Discharge Visualization **(GDV)** technique. This technique allows studying photon and electron emission in the electric field of various subjects- from the human being to precious stones.

A patient places a finger on the glass electrode of the instrument, a slight buzz rings out, and a blue glow appears around the finger. No pain, no feelings. A special TV system converts the glow into a digital signal, and a dynamic, live picture of flu-

Fig.5. Congress "Science. Information. Spirit."

orescence appears in the computer screen. Then modern software comes into play, extracting information from these files.

Processing is based on techniques of artificial intelligence and nonlinear mathematics. In short, a usual intrusion of mathematicians and technicians into a shaky world of human sciences. And quite a successful intrusion. All this was born in Russia, but as a rule was first valued abroad. Presentations in the majority of capitals of Europe; seminars all over America and in many countries of Asia; contracts with different companies; scientific projects in the universities of Europe, USA, National Institutes of Health of the USA. Then, gradually bioelectrography found acceptance in the native country. The congress - a culmination of development during the year, conclusions and future perspectives. The main direction of scientific congresses - medicine. Many scientists understand the limitations of modern medical knowledge and methods. Physicians and researchers had been delved deeper and deeper into subtle mechanisms of life, into the work of particular systems and organs, but then at some stage they found that the notions on the work of all these systems as a single whole are rather rough, or even primitive. An organism is a clockwork, chemical boiler, electrical machine - these are the main milestones. But a human being is not a machine, not merely a genetic code. It is a material body, plus soul and plus spirit. How to consider all this? And, moreover, measure? The EPC bioelectrography technique turned out to be exceptionally useful for these means. Imagine: a pregnant woman comes for a consultation and the first thing she does as she entered - places a forefinger on the instrument's window. In 20 seconds the display shows a number which notifies about the danger of unfavorable course of pregnancy with more than 95% probability. If the number is less than 6 - everything is OK, welcome; from 6 to 9 - please, follow an examination with a doctor, more than 9 - promptly! to the emergency ward. Fantasy? No, this is a technique developed and tested by doctor V.Gimbut and colleagues. Analysis of degree of seriousness of bronchial asthma, of state of gastrointestinal tract, spinal column, and all these in interaction with other systems and organs. Monitoring of condition of surgical patients after the operation, oncological patients in the process of therapy, estimation of subtle effects of acupuncture, classical and homeopathic medications. Easy, noninvasive, and relatively cheap.

And here a reader-Skeptic appears on the stage. There are many of them all over the world, not only in our country. They know everything, are firmly convinced in everything, and regard new ideas as an impingement on the existing system. Previously, on the social system, now, only on the way of thinking. Usually they wear black tie, vigorously reject humor! (On the whole, humor is the highest achievement of the human mind. None of the animals, even the most clever dog or an elephant can not understand humor. This is purely human, and requires a very developed intellect. What is more, humor is national. Try to tell our Russian anecdotes to foreigners - hard and thankless thing. Same as for us it is a serious intellectual test to understand English humor). So, let us put the Skeptic on the pages of this book and argue with him a little bit, at any rate answer his questions. We often listen to them in different halls of different countries of the world.

Skeptic. What you have told is interesting, but sounds like fantasy. I haven't seen such instruments.

Author. The instrument passed serious clinical tests in the leading scientific institutions of Russia and is certified by the Russian Ministry of Health as a medical instrument. And it is obvious that you haven't heard of it. The technique is still "young" and has become a property of a wide range of specialists and physicians only recently. Apart from medicine such topics as "INVESTIGATION OF PSYCHIC ENERGY OF HUMAN", "DIAGNOSTICS OF PSYCHO-EMOTIONAL STATE"; "STATES OF CONSCIOUSNESS: BIOELECTROGRAPHIC CORRELATES AND ENERGO-INFORMATIONAL MECHANISMS", "INFORMATION FIELD SCHEME OF LIFE", as well as a whole series of similar directions concerning research of consciousness, both in the process of normal vital activity and in special altered states, have been discussed at the congress.

Skeptic. Yes, we know, this is your bioenergetics. People have nothing to do, so they study informational field. It was good when scientists were formerly sent to dig potatoes. Was there at least some benefit from your activities? And what benefits do we have from your research?

Author. This line of research appeared to be very practical. Many years ago Prof. Pavel Bundzen, a Science Director of St. Petersburg Research Institute of Sport, decided to answer the question: what is the difference between top sportsmen -

Olympic and World Champions - and sport activists? The entire arsenal of modern scientific techniques, from electrophysiology to genetics, was used in the research, but real success came only when Prof. Bundzen started to apply bioelectrography. At the present moment the State Committee of Physical Culture and Sports has approved the "technique of EPC bioelectrographic diagnosing of the competition readiness of athletes" as a government approach and introduced it in the main Russian Olympic training centers. Later on we will discuss this topic, same as terrorism, but this is a special issue.

Skeptic. Are you saying that judging from the glow you can determine if this athlete is good or not?

Author. Yes, exactly. And also if the sportsman is ready for the competition at certain time or not and what are the shortcomings of his training. This was found to be a very effective and gainful approach. And what about the topic "ENERGOINFORMATIONAL TECHNOLOGIES IN AGING RESEARCH"? Another line of research. The percentage of elderly people in our country increases. Soon there will be one pensioner per one working person. Therefore, it is important to help elderly people to maintain activity, capacity for work, and diagnose all disorders at an early stage.

Skeptic. Of course, and then push through the law increasing the retiring age. Say that these glows become brighter after death, and the tax police will start asking income statements from the other world.

Author. I have no idea about the tax police, but "LIGHT AFTER LIFE" is one of the directions of our research, and a book of this topic has just recently been published for the third time. This topic was also present at the Congresses, for example, "REINCARNATION. CONNECTION WITH STORIES BY PEOPLE WHO PASSED APPARENT DEATH". Not only people can glow, but also many other subjects, even non-biological. However, in contrast to the human being, their glow is quite stable in time. Many subtle aspects of functioning of a subject can be discovered from this glow. For example, there is much talk about live and dead water. Is there some sense in this talk? It turned out that there is. Both theoretical approaches to the investigation of water, for instance "FLICKERING FLAME IN WATER GIVES BIRTH TO LIFE AND SUPPORTS IT", and purely practical - from parameters of fluorescence of water in different conditions to the system structures, increasing water energy level are discussed at the congresses.

Naturally, if it is possible to investigate water, we are fascinated to see the glow of blood. Very interesting results were obtained, which can become the basis of new clinical method. Another important task: can natural oil be distinguished from the synthetic having the same chemical contents? Standard methods of research give no opportunity to solve this task. The difference can be clearly determined from glow. Then we can find how various oils influence human states.

"And is it possible to study the glow of precious stones?", this question was once put by a jeweler from Australia.

"Of course, if we had subjects to study", I replied.

Next morning he came again, took out a small packet from his pocket, and strew a handful of sparkling diamonds all over the table.

"Here, this is for the first experiments. Later I can bring more", he said.

"Fine, thank you. Do you need some receipt?"

"No, I don't. Put them on the velvet ribbon, so that they don't roll away."

From this we started our experiments with stones. Then an Indian physician, doctor Shah came to Russia. He treated patients in Bombay tying precious stones to their acupuncture points. Professor Lev Kukuy organized a clinical test of this technique for the patients with cardiac diseases at the Pokrovsky hospital. Further research proved that mineral therapy really gave positive effect. And the study of energy state of stones became a usual topic of bioelectrographic research.

Skeptic. Well, started on a merry note, but finish on a sad one. Medicine, psychology, these are respected spheres. But about stones - this smacks of chiromancy. Maybe you also studied healers and extrasensory people?

Author. Of course. This is a very interesting topic. The easiest would be to declare that this is charlatanism, to call it pseudo-science, and head a "Quack Busters"22. Fortunately, less and less depends on rigid old-minded scientific committees now. So, the healer Alan Chumak and other well-known individuals attend our Congresses, and the character of glow demonstrates if a person can do something or not. We can discuss it in more detail.

Skeptic. Listen, what you named are topics for the work of a whole institute. And seems you have only a small laboratory in St. Petersburg. Who runs all these directions of research?

Author. Yes, we have a small laboratory, and most of the above mentioned research was performed by specialists from different institutions and universities in different countries.

Skeptic. Another shady business.

But, it seems that the reader is already tired of boring enumeration. Let me better tell you a few more real stories, which took place in recent years in different countries and situations. Then you will better understand what we are really doing. At the same time we will speak about healers, extrasensory people, telepathy, holy water and vision without eyes.

A COMPUTERIZED AURA

I have come into the world as light.
Bible

Once we held a workshop in a little northern city Oulu - a scientific and industrial center of Finland. After the lecture I was approached by a young man who asked for a consultation. He was strong and well-nourished, but the eyes harbored anxiety, what is more the behavior was somewhat shy and diffident.

"I have certain problems, maybe you can advise something. I have done dozens of analyses, but they can find nothing."

"Fine, let us take a picture of your field," I agreed and turned on the computer. The young man put the finger on the glass electrode, a short buzz rang out, and...

"Another finger, please."

No unpleasant sensations, no pricking. All the procedure with his finger took half a second. The Brave Finnish young man had no time to be frightened. We took pictures of the лишний артикль their fingers for another two minutes: changing fingers was an adequate task for the young man, but required time. Processing and analysis of data in computer took another 10 minutes, after which I said,

"Well, let us discuss your situation. Apparently, you have something like chronic fatigue. Depressed state, blue study, you don't want to do anything. No energy and forces for that. The doctors do not find anything and say that you are physically absolutely healthy. The boss thinks you are a malingerer. Though does not talk up about that. Everything goes amiss in life. And everything started from the financial problems which you took to heart and felt keenly. Before that you worked very tensely, forgetting about food and saving on sleep. You have never been attracted with bars and discos, you wanted something serious, and therefore you put all your efforts into the work, but failed in your first independent business, moreover because of the betrayal of your best friend. These all together caused deep depression, which has now developed into diseases. You don't know what to do and nobody can help."

I ceased talking and looked at the young man. He seemed to be completely confused, even his mouth was half-open in amazement - a typical classical picture.

"How do you know?" he forced himself to speak, "Somebody of my relatives told you?"

"No, my friend, I see you for the first time in my life. ("And, hope, last time," I thought to myself). "All this is written in the picture of your field, and I just analyzed this information."

"And what shall I do?" he asked perplexedly.

"You need to get out of it. Imagine that you fell into a pit. Now you have to take your legs and pull yourself out."

"But how? I tried both exercises and vitamins, but no result."

"First of all, look at yourself from the outside. Come out of your body and situation. Understand that all your problems are in your attitude to yourself and your life. Fall asleep today and wake up tomorrow in a new life, where there is no regret for what was. The past has already gone. Return the painful thoughts to that old man, and start a new page. By the way, you have no problems on the physical level."

"But I am sometimes sick."

"Don't drink coffee with milk and, generally, pay attention to the selection of food. You have weak stomach. Drink 2-3 glasses of red dry wine and less coffee."

"And I drink 5-6 cups a day. We always have a thermos bottle with hot coffee at our company. As a matter of fact, there is such a habit in Finland. The Finish overtake Brazil in drinking coffee and play hockey practically the same as the Brazilians play football."

We talked with the guy for another half an hour and he left slightly braced up. It is still pleasant when you are told that no fatal illness is found and "most diseases are rooted in one's nervous system." But in reality it is not an easy matter to overcome these diseases! Many doctors suspect that psychosomatics is one of the essential causes of serious organic diseases, from diabetes to oncology.

How did I manage to know all this information about the Finnish guy I had never met before? To make everything clear: I am not an extrasensory individual and not a fortune-teller by stars or hands. We receive all information about the state of people using computer analysis of fingertips glow. I will later explain what it means, but now I just reproduce the line of my argument.

Fig. 6 shows the picture of field of this Finnish guy. See how weak it is, with many breaks and "holes". And now, fig. 7 - his field, but taken with a special filter. Absolutely different picture! Powerful and bright field! No holes, no breaks. The first picture demonstrates psychological field, the second - physical. Consequently, the first conclusion: this guy has no problems on the physical level, but the nerves are out of order. What is more, these are not psychic disorders: in these cases the picture of field looks differently. Disorders shown in the diagram (Fig. 8) in the stomach area indicate that there is a slight energy deficiency in this area - potential physical weakness. With optimal nutrition and way of life one might never recall this. Unless he eats fried pastry, French fries, has a thermos with coffee on the table and, what is more, is nervous - ulcer is guaranteed.

Now let us look at the picture of fingers' glow, EPC images, as we call these pictures (Fig. 9)

The left side is significantly weaker than right. The left hand is refers to the right cerebral hemisphere, the right one - with the left. The right brain is responsible for emotions, feeling of anxiety, intuition. The left brain - logic, speech, planning, and optimism. It is also said that the right brain is female, and the left - male. In other words, all the best in human, delicate, tender, and heartfelt comes from the right brain, and all rational, provident, and strict - from the left. Both hemispheres interact, exchange information, what is more, in normal conditions the left brain dominates. It is said that both hemispheres live as an old married couple, they both have a stable attitude to life, understand each other and can get along well.

Each finger is related to certain systems and organs. This is based on the ideas of traditional Chinese medicine about the energy channels, penetrating through all the human body, but we will discuss it later. Now the important thing for us is that the glow of fingers is not equal. Note: the first finger has quite bright fluorescence, the forefinger and middle finger - much less, and the ring and little fingers - again strong glow.

Here comes the line of reasoning.

Energetics of the thumb is strong both on the right and left hands. This is an indication of active mental work with a tendency to logical reasoning. This conclusion is proved by the diagram of the chakra energy level (Fig. 10, 11)

The upper chakras are active, then a slowdown comes, and again a rise to a lower chakra.

Imagination and emotions play a significant role in the mental work of Pekka (this was the guy's name (strong glow of 1L finger). Such people are usually inventive and are inclined to invent a lot of new ideas and plans, but aren't always able to bring them into life. However, the right brain also manifests anxiety and pessimism. Therefore, creative intellectuals with developed right brain often fall into a depression. They often lower their hands at the very first failures or signs of resistance to their wonderful plans. Therefore, depressed state of Pekka is explained by the type of his energy field.

How did I manage to disclose that he had financial problems? Very easy. There are only two topics for a young 28-year old man he is deeply interested in: business and women. The energy level of Pekka's sexual areas is quite active, although unbalanced. This is expressed in the field picture taken with filter (Fig.12)

Do you see a typical outburst in the lower part of stomach? In addition, the diagram (Fig. 7) shows high values for both hands in these spheres, which is the evidence of functional balance and active energy state. I'd like to remind that for estimation of physical state we analyze data obtained with using filter, i.e. the outer diagram. The Inner diagram (pictures taken without filter) demonstrates psychoemotional status. Therefore, if not women, then finance. This guy had problems in business relations.

And, finally, the last argument in the line of my reasoning was the data obtained in "EPC Activation" program. This program combines results of measurements of fingers' glow with and without filter and calculates a few coefficients (Fig.13).

As we see, "The level of stress is very high A=9.940!!!" In a 10-point scale this is practically the highest limit. That's the end! At the same time, the health factor for both hands corresponds to the norm, i.e. it is a practically healthy young man in the state of strong distress.

So, as you see this was not a fortune-telling, but strict mathematical analysis with the calculation of quantitative indices. What is more, the work of all programs took a few minutes, and talk with the patient - half an hour. It is necessary not just to tell him how bad everything is, but try to find the source of the problem. It can be on the psy-

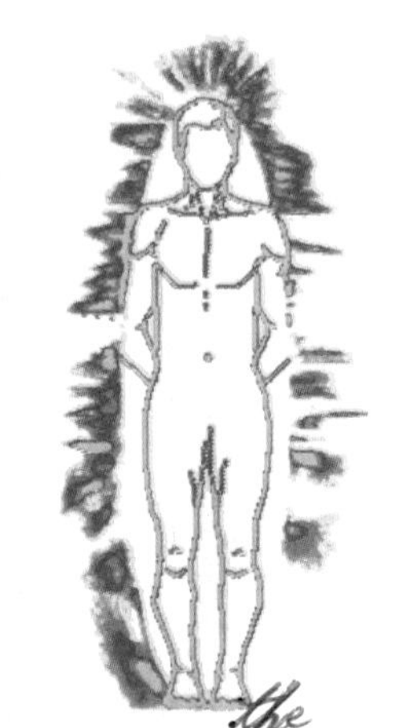

Fig.6. The picture of energy field of a Finnish guy.

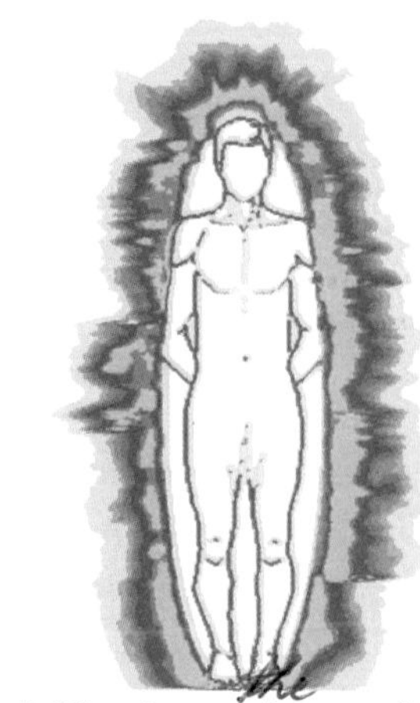

Fig.7. The picture of energy field of a Finnish guy taken with a filter.

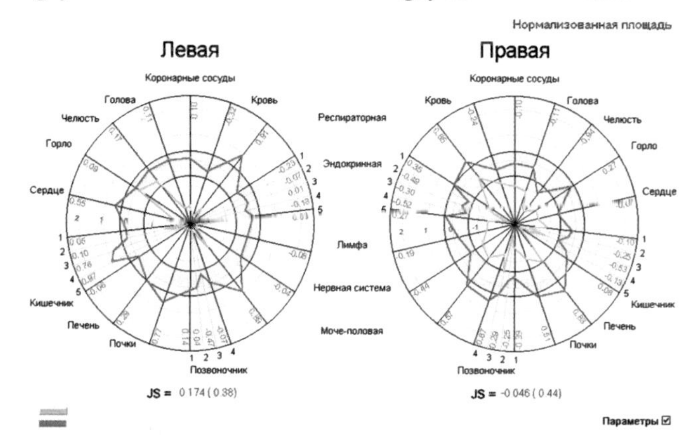

Fig.8. EPC Diagram of a Finnish guy.

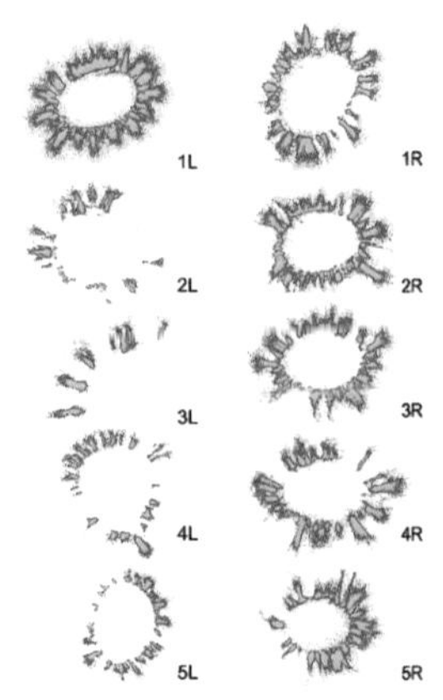

Fig.9. EPC images of fingers of a Finnish guy.

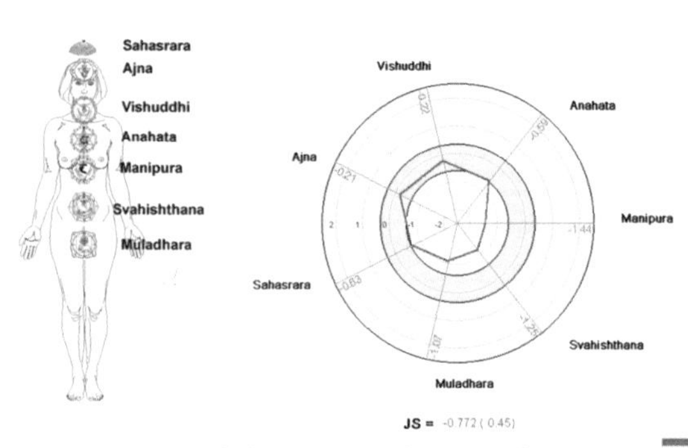

Fig.10. Chakra energy of a Finnish guy.

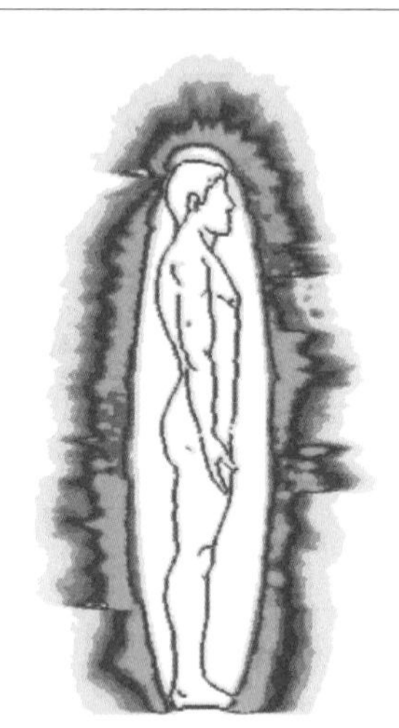

Fig.12. The picture of energy field of a Finnish guy taken with a filter.

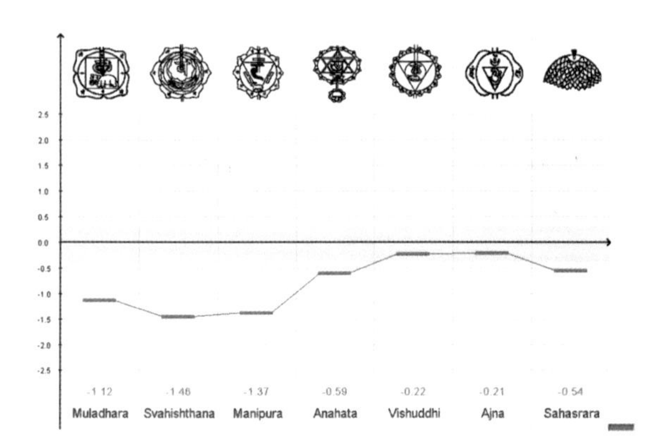

Fig.11. Chakra energy of a Finnish guy.

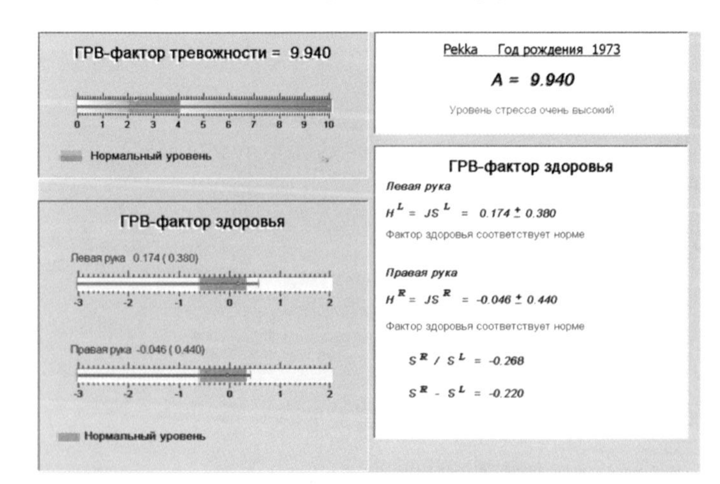

Fig.13. Stress level of a Finnish guy.

chological level, as in the case with Pekka, but can also lie in physical problems. What is more, a person observes the symptoms and starts visiting doctors, but by no means do the symptoms always show the real reason of the state. It aches in one place, and the source of the problem is somewhere else. This fully pertains to such complex diseases as hypertension, diabetes, migraines, and many other ones. And here is the EPC technique, estimating the state as a single whole, demonstrating complex, systemic and holistic picture of the functional state of systems and organs, gives its indispensable help. One does not need to sit in waiting-lines at the consulting-rooms of various specialists, but directly works on the systems requiring special attention. The finger glow analysis with the EPC method gradually steps into official medicine as the first stage of express-diagnostics. Why gradually? We will discuss this a little bit later.

Skeptic. This might be easy and understandable for you. But a normal person will never master this.

Author. You shouldn't think so. The most important is to be able to press computer buttons and pass a training course. In two days everybody starts working on his/her own, although, of course, real understanding comes with the experience of practical work. People who have basic knowledge in Traditional Chinese Medicine can do all this much easier.

Skeptic. I thought so. You have everything based on Chinese and Indian ideas, but they have not yet been proven. All this is a fantasy.

Author. These ideas are several thousand years old and they are still alive. And what's most important, they help millions of people to be healthy. You know, there is a wonderful proverb: "You can fool one person for a long time, and fool all people for some time, but you can not fool all people for a long time". So, the first criterion of truth of Orient ideas is practice. But now there are a lot of scientific research works, proving their truth.

Skeptic. But why from finger? Are there no other places on the body? And, generally speaking, where did you get it from? Suddenly, out of nowhere?

Author. Certainly, not. Investigation of "live glow" has a long history, and we can devote a few next pages to it.

ELECTROPHOTOSPHENES AND ENERGOGRAPHY

Allah is the Light of heavens and the earth.
Koran.

"Electrophotosphenes and energography as a proof of existence of the physiological polar energy". This was the name of a small book by a doctor from St. Petersburg, Messira Pogorelsky, where he described his experiments in bioelectrography. Book, published in 1893. Many photographs of the glow of fingers and toes, ears and nose show how the pattern of fluorescence varies when the psychic state of a person changes. However, this work was far from being the first one.

In the 1770's many researchers made experiments with electricity. There was no practical benefit from this: more than 100 years still remained till the invention of the electrical bulb by Tomas Edison; life passed with candle lights, European monarchies seemed to be eternal, and science studies were mainly the lot of an aristocracy.

In 1777 a German physicist George Lichtenberg touched a metal electrode covered with glass and connected to voltage with his finger while experimenting with the electrical machine. And suddenly a burst of sparkles flew all around. This was magically beautiful, although a little bit frightening. Lichtenberg jerked back the finger and then repeated the experiment. The finger placed on the electrode was shining with bright blue light and treelike sparkles dispersed from it.

Lichtenberg, being a real academic scientist, investigated the behavior of this fluorescence in detail, although he substituted a grounded wire for a finger. The effect was the same, which later suggested an idea that some special energy exists in the body, and first electrical then torsion properties were attributed to it.

Articles by Lichtenberg, masterfully done in German, are still cited in books on gas discharge. Further research demonstrated that electrical fluorescence was rather common in nature...

We were climbing the summit with a group of rated sportsmen in the Dombay region of the Caucasian mountains. Early in the morning, before daybreak, we climbed the glacier with perilously beetling ice lumps, and with the first light of day we were on the rocks. Every hour we climbed higher and higher, driving in the titan pitons, hanging up ropes, and the ice glacier lumps gradually turned into small black points below. By the middle of the day the sky was covered with clouds, hanging low overhead ready to rain down. Our climb was planned for a day: we intended to reach the summit by 4 - 5 p.m. and then come down before darkness. Thus we could go with light luggage having taken only equipment and some provisions. Indeed, we climbed the ridge by 5 p.m. That was a narrow stone spine, breaking with steep walls on both sides. Half an hour accurately balancing on the ridge - and we were at the top. The lads cheered up and started discussing the menu of the evening dinner in the camp. I anxiously looked at the impending clouds. Suddenly we heard a thin clinking sound, like a violin string.

"Look, guys, my ice-axe is shining!" exclaimed one of the mountaineers. The end of his ice-axe turned to our side blazing with a bluish glow.

"Throw the ice-axe right now", I cried. "Guys, put all iron equipment down and pile up. Now! This is very dangerous!"

We started putting off the carbines, climbing irons, hammers, and soon it was a large iron pile on the ridge site. The iron pile droned and shined with bluish light.

One of the alpinists, Sasha Strelnikov, stepped up to the pile and wanted to put his ice-axe together with other equipment. Suddenly a dry crackle rang out; he awkwardly stretched his hands and rolled down the mountainside.

"Hold him," I shouted wildly, and together with a few guys jumped to the rope slipping away into the precipice snakelike. Fortunately, all of us were bound to one another with the ropes, and after a strong jerk, everybody fell down, all our group froze in the very odd poses having clutched at the nearby stones. We carefully pulled Sasha out and back to the ridge: he was in his senses, but couldn't move either a hand or leg.

"What was that?" he asked in a weak voice.

"Atmospheric electricity. Something like St. Elmo's Fire. The sharp beak of the ice-axe created a disturbance in the atmosphere and provoked a lightning discharge. You were lucky that this discharge was not very strong."

"What will be with me?" plaintively asked Sasha.

"Lightning is Jupiter's tool. Now you have his mark. So we will have to carry you down as a special value to humanity. Maybe you will start to read thoughts or predict the future. We will check you with our instrument when we come down."

In half an hour, having drunk hot tea, Sanya could walk on his own, but all his movements were slow, uncertain, and the three alpinists had to look after his every step using the ropes. The night caught us at the rocks. We spent it sitting on narrow shelves, our teeth chattering and telling political anecdotes. We reached the base camp only in the middle of the next day, where the longed-for roasted chicken was waiting for us.

I don't know if Sasha showed extrasensory capabilities, although this often happens after a lightning stroke. But it was found long ago that in the prestorm environment, when the air is filled with electricity, many sharp subjects and often the human body start glowing. Is it the reason for nimbuses and aureoles of saints?

In the Nineteenth century enigmas of electricity were opening to people. One of the great inventors was Nicola Tesla, from whom we now have lamps and television sets. He invented the generator of alternating current. However, if it had not been him, somebody else would have done it. Inventions come to life when a social need for them appears. Then different people simultaneously and independently start arriving at the same ideas. This is connected with the fact that the ideas have their logic of development, and the developers shall only intuitively feel this logic.

After raising good money with his patents, Nicola Tesla began the mysterious experiments on energy transfer without wires. He did not finish his developments and died in destitution, but up to now enthusiasts have been trying to investigate his ideas. We get used to our technical progress and reap its fruits with pleasure, but is it the only possible way of development?

At the peak of his career Tesla liked to give public lectures and impress the audience with the following experience. The light was turned off in the room, Tesla turned on the generator of his own design, stood on the platform-electrode, and his body got wrapped in the glow. The hair stood on end, glowing rays of light radiated in the space. The experiment was very effective, though not all those who wished managed to repeat it: as a matter of fact, their glow was much less and for some people even missing. Not in vain was it said that Nicola Tesla had special energy state.

Further research did not go much beyond investigations of the glow of fingers, sometimes ears, nose and other prominent parts of the body. Is it possible to reproduce Tesla's experiments and make all the body glow? Yes, it is. But is it necessary? Powerful equipment, which is not safe if not handled properly, is required for such an experiment. Moreover, the stronger electrical glow, the more ozone is generated in the air. A high concentration of ozone is far from being healthy.

So where is the similarity in the experiments of Lichtenberg, Tesla and the glow of the ice-axe? In all those cases the gas discharge appears near the earth rod. High field intensity is formed near its sharp end when placed into an electrical field. Electrons, which always exist in the air or are emitted by the bodies, start speeding up in this field and, having picked up necessary speed, ionize air molecules. Those, in their turn, emit photons, mostly in the blue and ultraviolet spectral regions. Here the glow appears. What is more, from the viewpoint of physics both a nail, an ice-axe, a human finger, and a person can be the rod. Everything depends on the scale.

"But this is certainly dangerous", exclaims the **Skeptic** appearing from nowhere. "You said yourself how your climber was conked, and now suggest everybody should tread in his steps! Should there be natural selection to survive after the electric chair?"

"My friend, several laws of physics and physiology work to our benefit. Not electric voltage, but electric current is dangerous for the human. Current, which can flow through the body, depends on the power of its source, frequency and skin resistance. For example, power in an electrical socket is limited only with a safety-lock, it usually amounts to 5-10 kw. Skin resistance of different people changes from thousands of Ohms to dozens of millions of Ohms. So, it turns out that putting fingers into a socket with the mains voltage 110 V, one can get current rush from hundreds of milli-Amperes to tens of micro-Amperes. In one case it is practically lethal, in the other - the person will not feel anything."

I have known an electrician who cleaned the wires alive with his teeth. His skin was as the skin of a hippopotamus.

Wet skin has much less resistance, as shown by examples of electrical devices thrown into the baths with villains in American thrillers.

Generators used in Bioelectrography have very small power. It means that they can not give high current, even if you lick the electrode with your tongue. In addition, these generators make use of high-frequency voltages and short impulses, and by the laws of physiology such current can not penetrate into the organism, as it slides on the skin surface.

And, between you and me and a lamppost, electricity is a dangerous force, which should be treated very carefully. For example, microwave ovens.

"Well, that's a bit too thick! They are used world-wide and nothing happens", says the **Skeptic** with offense and not having said good-bye, disappears.

PROFESSOR OF ELECTROGRAPHY, BRAZILIAN MONK AND SOVIET ELECTRICIAN

God being truth, is the one Light of all.
Adi Granih

The interest in photographing electrical fluorescence arose all over the world after the experiments of Nicola Tesla. It took two evenings to assemble a Tesla generator and start the experiments. From the middle of the Nineteenth century, the glow was first registered on photo plates and photo films.

And, when the images started to be registered and not just admired, it was found that the picture of fingertips' glow depended of the subject. Someone felt nervous or, on the contrary, fell into a meditative trance, and the photo of glow changed its form.

A significant contribution to the study of these photographs was made by a talented Byelorussian scientist Jacob Narkevich-Yodko in the end of the Nineteenth century. He was an independent landowner and spent most of his time on his estate above the river Neman. There he actively experimented with electricity, applying it in agriculture and medicine. A straight parallel with modern medicine can be drawn from the description of experiments on the stimulation of plants with electrical current, on electrotherapy, and magnetism by J. Narkevich-Yodko. But the scientific achievements of our time are not just "the new as a well lost old". This is a new convolution of perception. In the end of the Nineteenth century, when the principles of electricity were only emerging, when the main source of light was a kerosene lamp, the searching investigators were trying to apply electricity to the most different areas of life. They were as if naming the chapters of a new book, but had not enough power to write the contents. Therefore, we find the sources of one or another modern scientific direction in the works of the enthusiasts of the Nineteenth century.

J. Narkevich-Yodko developed his own original technique for making electrophotographs. He made more than 1500 photographs of fingers of different people, plant leaves, grain, and in the 1890's this research attracted attention in the scientific community. In 1892 J. Narkevich-Yodko presented for the professors of St. Petersburg Institute of Experimental Medicine, after which he was appointed a "Member-Employee of this Institute" by the order of the Institute patron, the Prince of Oldenburg. The results of Narkevich-Yodko created such an impression upon the scientific community that in 1893 a conference on electrography and electrophysiology was organized in St. Petersburg University. In the same year Narkevich-Yodko visited the scientific centers of Europe: Berlin, Vienna, Paris, Prague, Florence and gave lectures there. His experiments on electrography were acknowledged as important and envisaging further development everywhere. Narkevich Yodko received medals at several exhibitions, and at the Congress in France in 1900 he was nominated a professor of electrography and magnetism.

J. Narkevich-Yodko combined scientific work with public activity. He organized a health center on his estate and received people from very different social circles: from grandees to plain people, and cured many different diseases with the help of the newest for those days, methods. But with the death of J. Narkevich-Yodko the contemporaries forgot his works. He made an interesting discovery, but could not overcome the barrier which had always been in the road of wide introduction. There had been so many interesting methods, inventions, developments, which disappeared together with their authors! In order to make an idea publicly acknowledged it is necessary to introduce it deep into the collective consciousness, attract students, followers, and companions. It is obligatory to publish articles, books, written by different authors and, advisably, in different countries. Various researchers should independently test a new idea and make sure that it is effective, but in order to do so they should have a desire! And such "overcoming of a potential barrier" usually requires more than ten years. If the author has enough persistence, energy, and optimism to get his own way, the idea starts living independently and sometimes the author gets the interest. If not, then not. And, naturally, much depends on the favor of "lady Luck".

The general social situation in Russia played a role in the life of Narkevich-Yodko. Evil social winds were blowing, which turned into a hurricane having destroyed the leisurely way of life of the Nineteenth century and changed the beautiful estates above the Neman and Volga rivers into cold abandoned ruins. "No prophet is accepted in his own country", especially when this prophet thinks in a non-standard way and doesn't fall into the usual pattern. But can a Prophet live a normal life?

At practically the same time, on the other side of the globe in Brazil, very similar experiments were performed by a Catholic monk, padre Landell de Morua. This was a funny little man with a long nose, disappointed in the vanity of the worldly life - and bound to devote himself to serving God. A monk's life left a lot of free time, after reading prayers and performing rituals. Some of the monks went in for gardening and sometimes, as Gregor Mendel, invented new laws of nature; somebody else researched ancient civilizations, which were all over the place in South America; but padre de Morua started inventing. He invented the technique of photoregistration of electrical glow and started giving lectures and writing to social leaders in order to attract attention to his offspring. But Brazil is not the USA. In Brazil everybody enjoys life, dances, prepares for the carnival for half a year and continuously sorts out emotional relationships. Have you watched Brazilian soap operas? They live like that in reality. Well, probably, not so many intrigues. It is too hot there in summer to trouble oneself with much effort. I had a chance to be in Brazil at conferences, I will tell later about that, and every time it was a Holiday with a plenty of food, wine, songs, and dances. Therefore, it is no wonder that the invention of padre de Morua produced much rapt attention, congratulations, banquets, but was not widespread. Then the little big-nose priest invented the radio (practically simultaneously with

Popov and Markoni), but again he was unable to draw in large crowds. Even the military. Generals admired, realized the importance and perspectives of the wireless communication, promised to call the colonel the next day and assign resources, but in an hour left for a night's banquet and forgot about everything. It is worth mentioning that after more than 100 years the habits of South American generals haven't changed a bit, which I witnessed myself. When you have a date at 5 p.m. and the person comes at 8 p.m., this is normal. If he comes the next day - this is excusable. If he doesn't come at all and doesn't call, and meeting you in a week says,

"I am so sorry, our meeting did not come out last time. What about tomorrow 5 p.m.?" this is typical.

So, padre de Morua was creating interesting devices all his life and trusted God that he would attract somebody's attention to the padre's inventions. God answered his prayers, but time passes slowly in heavens, and the interest to the padre's work has arisen only now. Brazilian historians disclosed many small details from the life of the little priest and published several books. At the Congress in Brazil I was solemnly presented with the bust of padre de Morua, and a pretty Brazilian woman, sitting next to me at a banquet, was telling about his wonderful inventions for a long time. I felt sorry that I didn't understand anything in Portuguese - I could have learned a lot of interesting facts.

J. Narkevich-Yodko had the same fortune in Russia as padre Landell de Morua in Brazil, as well as many other researchers all over the world. Their works acquired interest, were acknowledged, but they did not have enough organizational and business skills in order to overcome the "infancy border" of their inventions. A first step was made, but it was not enough. Another half a century was required for the further steps.

In the beginning of the Twentieth century nobody even recalled the mysterious glow. There were many other problems: wars, revolutions, breakthroughs in physics, discovery of antibiotics and roentgen rays - everybody was sure that it was very close to the outright victory. However, this victory was understood in one's own way and anyone who was against this understanding was to be destroyed. Only by 1930ies the life more or less came right. And here appeared the mysterious glow again. And, as if by chance, it was discovered anew, but there is a rule behind every chance.

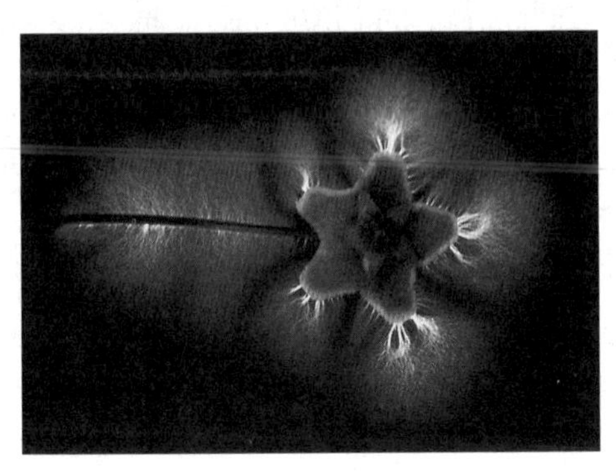

THE KIRLIANS

I have been developing lines of research connected with the Kirlian effect for several years already. Therefore, I would like to say a few words about these wonderful people.

In 1978 we got an invitation to participate in a conference dedicated to the 80-year anniversary of Semyon D. Kirlian. The conference was held in Krasnodar city, in the inventor's home land. It was organized under the aegis of "Quant" - a big research and production center, which was working on all questions connected with the sources of supply in the USSR: from clock accumulators to solar batteries. The central institute was located in Moscow, there was a large department in Krasnodar, headed by A. Skokov - a talented organizer, who later managed all research-production organizations and moved up to the position of Gorbachev's adviser. When the big international interest in the work of the married couple Kirlians was noticed at center "Quant" and benevolence of the academic authorities was shown, Semyon Kirlian was provided with a small room in the institute and several colleagues for assistance. The colleagues determined that the effects observed were real and reported about that to Moscow. And the authorities decided to make their name in the Kirlian effect. A scientific topic was opened, a popular science film was made, and a conference was organized. Owing to that work several colleagues obtained apartments in Moscow. However, in a couple of years it turned out that the first cream had already been skimmed off the milk, and in order to get more serious implementation long and laborious work was required. And that was the task of the academic organizations and higher education institutions, technological organizations, even the big ones, were not adapted for such work. Therefore, the topic was closed, the colleagues were reoriented, and the Krasnodar laboratory had existed on the enthusiasm of a few colleagues for many years after Semyon Kirlian's death, and finally became part of a medical center.

Of course, our main impression in 1978 was the meeting with Kirlian. He was alone that time - his wife Valentina had passed away in 1971. It was an old man, modest and not tall by appearance. His behavior showed that the walls of the home laboratory were more customary for him than the honorable chair at the scientific meeting. But apart from that it was understood that he knew his place and took such an impressive meeting confined to his anniversary as a matter of course. Deep intelligence and interest in everything which took place was felt in his slightly prominent lively eyes. He didn't take part in the scientific meetings apart from a short welcoming speech, it seemed that all those academic discussions were strange for him, but when we came to his home, a small apartment in a five-storeyed gloomy-looking building, he bucked up and showed us around his home laboratory with pleasure. It was obvious that he got used to a flow of most eminent visitors and all that fuss wasn't creating a big impression upon him. All his appearance was like saying: "I have done my part, now it's your turn!"

Semyon Kirlian spent most part of his life with his wife Valentina in a poor two-room apartment at the corner of Gorky and Kirov streets in Krasnodar. The wooden two-story house where they had started their family life was swept away by progress - a building program turned the small provincial town on the banks of the Kuban river into an industrial center. Now Gorky street is covered with brick five-storeyed structures, one of which has a memorial plaque in commemoration of the married couple Kirlians. He was an electrician, she - a school teacher. He had the practical mind of an inborn inventor, which brought him the honorary title "Honored Inventor of the USSR" and world fame afterwards. She was an intellectual, a university graduate, and his apprentice.

She was giving literature classes at school during the day, but the rest of Valentina's time belonged to him. She was his Love, his Passion, and his apprentice. She helped him in making experiments, which took the whole evenings, and often whole nights, all weekends, and all holidays. They were deeply carried away with the experiments with auras of live subjects, and since 1939 they had worked hard. The only rest they could afford was walking hand in hand under the trees and along blossoming fields so typical of the South Russian cities.

The main property of the Kirlians was their laboratory. Their small bedroom was filled with equipment. Every evening in order to go to bed they had to put away the photoplates, developing dishes, and induction coils from their bed. The black monster in the corner was the high-voltage Tesla generator assembled by their own hands. At night they put cloths on it. And in that room Valentina passed away in December 1971.

Semyon cut the gravestone himself and set it on the grave with his friends. Within a month after Valentina's death he turned into a crooked old man. He had never been religious, but his experiments gave him belief in the life after death. As he explained, he realized that watching the last flashes of glow of the dying leaf. He believed that "bioenergy, which produces the vibrating fluorescence of biological subjects, will never deplete, even when it leaves the dying body in the last flash of fire - it goes away into space". Consequently, the soul is eternal. And he engraved a bunch of lilies with the aura around them on her gravestone - not just for himself, but for us, in order to remind about her and him, when his time would come.

Friendship with V. Krivorotov, a talented sensitive healer, strengthened his belief in that even more. They had spent many hours together, photographing fingers' glow in the process of healing.

When everything started half a century ago, Kirlian was a local master with skillful fingers, bringing to life everything he touched. Having only four classes of education, he could repair practically any electrical appliance, from burnt electric stone to telephone. He came with the tool case and a hank of wires at first requirement in order to repair wiring, regulate a junction box or electric generator. After a number of years he learned to repair photo cameras and microscopes. He invented what he didn't know. He drew his own schemes, and spent nights reading books. He loved electricity and all machines connected with it.

These were lean and violent years, when the USSR was getting out of the state of total chaos. Ekaterinodar city had been founded by Kuban Cossacks in 1794, who had named it in honor of the great Empress; the Bolsheviks had renamed it Krasnodar. When the new hospital was built and electrical equipment turned out to be non-functioning, Kirlian made it work. He could handle any equipment, from electro-massage to an X-ray unit. When his skills became well known, he started to be invited to all medical institutions of the city. In time a small two-room apartment was given to him and Valentina, where they had spent 40 years. However, at that time everyone in the USSR lived like that, and a separate apartment was considered to be a very good variant.

Once when he was repairing a high-voltage electro-massage unit at a hospital, a discharge came through his hands, but he didn't feel pain. On the contrary, he saw a very beautiful flash which was worth capturing. The idea inspired him. But how to take a picture of such phenomenon? At that time glass photographic plates were used and the captured subject needed to be pressed to the plate.

Well, and what about the subject? Of course, he needed to photograph himself. Which part? Why not a hand? So, everything started like that. He planned to use an isolating table and place the photographic plate wrapped into black paper on the surface of an electrode. The second electrode was gripped in hand, a finger was placed onto the plate, a click of a switch - and ready! He was standing on a rubber rug for isolation.

According to all responses, the original Kirlian equipment was primitively simple. He created it from the parts taken home from hospitals. Their main resources were spent for technical literature, which he bought and brought home. Valentine, being enamored of photography, developed and printed.

This is how the Kirlians described the discovery of the most interesting property of glow - dependence on psycho-physiological state 2:

"Perhaps, there has been no other device which would be as whimsical as our hand-made discharge-optical equipment. Moreover, it required triple precise tooling - optical, discharge and voltage tooling. The success of a demonstration fully depended on the experience and skills. It was impossible to keep calm.

As a rule, transparent cover and optics combined in one device was checked beforehand. The equipment was always demonstrated by one of the authors, observing his hand.

The demonstration starts. Putting the hand under the transparent electrode and looking into the ocular, the author turns on the generator with his foot. From the very first second the transparent electrode starts playing up, the background is not clear, and the channels have no brightness. The author asks the guest to excuse him. The electrode is quickly disassembled, everything is cleaned again, tooled, the generator is turned on - but the same unfocused picture in the ocular. The guest, tired of waiting, wishes to look through the ocular. And, how odd: he is as if pleased with the picture. As the guest gets interested with what he has seen, the author calms down, and meanwhile cries of astonishment are heard: apparently, the guest finds something exceptional.

Recalling that the working time is already over the author finishes the demonstration, looking into the ocular himself beforehand. He is more surprised than the guest: the visibility is excellent.

This unexpected case has become a whole revelation for us. Isn't it strange: five transparent electrodes suddenly and at the same time stopped working well with one particular object, or rather, subject, and suddenly unexpectedly worked properly with another one.

It was impossible to stand such an intriguing situation indifferently. The diseased author got up from his bed and we began checking the work of five devices together on one another. There were no doubts: electrical pictures of the diseased were chaotic, pictures of the healthy represented clear play of magical discharge flow.

This is where every cloud truly has a silver lining! This is what we found through quite an unhappy fact - vasospasm. The vasospasm brought about disorder in the observed picture of electrical state."

A question arises: how independent were the Kirlians in their work, didn't they take the very idea and the main techniques from their predecessors?

The reply is very simple. Even living in St. Petersburg it hasn't been easy to find information on the experiments of the Nineteenth century, and it is not clear how that could be accomplished in the provincial Krasnodar at all! But even if we assume that the magazine "Niva" from the beginning of the century, where photos of the glow were published, somehow fell into the hands of the Kirlians; in order to reproduce such photos it was necessary to pass the whole way from the beginning! Therefore, the Kirlians could be accused of various sins: lack of knowledge of English, lack of skill to dance rock'n'roll, but certainly not of plagiarism!

But their entire work could have passed unnoticed and become a property of enthusiasts - students of local lore, as it happened with many talented researchers and scientists. The Kirlians were lucky. Or rather not, "lucky" is not the right word. Their inventions appeared in the right moment when the basis was already laid. They acquired interest in the world. First of all, in America. Only then in the Soviet Union. And this interest made the name of Kirlian world-renowned and provided rapid development of their ideas. So, we can literally say that Semyon Kirlian lived a happy life. He devoted himself to the interesting work, he had his beloved wife who spent all her life with him side by side, and in the end of life his work was accorded really wide recognition. In 1978, on the 80-years anniversary of Kirlian, a scientific conference dedicated to the method of bioelectrography was organized in Krasnodar. Semyon Kirlian was decently sitting in presidium, silently listened to the presentations, did not break into the speeches, but it was obvious that he understood that the work of his life did not die, it was being developed further. I had a chance to meet him that time after the conference. In the close circle of associates he became cheerful, got into conversation, and discussed details of introduction of the "Kirlian effect" with pleasure. May God grant each of us life like that!

The Kirlians

BRAZIL, BRAZIL, THE COUNTRY OF SPIRITS

God is the sun beaming light everywhere.
African myths

November, 2000. The International Congress of the Bioelectrography Union in Brazil. Hot, summer, but unlike us Brazilians do not pay any attention to the heat. Many wore jackets, so I had to array myself in suit and tie. International meetings and conferences have certain laws: only high-hierarchy scientists like Nobel Prize winners have the right to wear jeans and T-shits, outstanding scientists can afford color shirts without jackets, jeans are considered bad form and can significantly compromise a reputation. It also depends on the country: it is indecent to come without a tie in Germany or England, in the USA it is easier: there you can meet a professor wearing a suit, but at the same time in jogging shoes.

Leading scientists, developing bioelectrography in the world came to this congress. Quite an imposing and interesting convention. What is more, many developed their own directions and brought devices which they constructed themselves.

Main dialogues and exchange of views take place in the evenings, after the end of sessions, with a glass of red wine. And so, when we were sitting and shared impressions and problems with one another, the President of the Finish department of the Union Matti Ollila suggested an idea:

"If all of us are here, why don't we test one and the same person tomorrow with different instruments? Make conclusions independently, pass them to someone from the Organizational Committee in the written form, and then compare the results. And, naturally, confront them with the real situation of the patient."

The idea was interesting. I happened to participate in such "blind" tests several times. Moreover, twice in a live broadcast. Of course, the risk exists, but on the other hand in case of success the trust in what you are doing grows significantly. Once it was in Dublin, at a popular night television show watched by all Ireland and part of England. After I correctly revealed problems of the analyzed people by the field picture, and they confirmed that in the hall right away, I was recognized in bars and stores for the five days while I was in Dublin. And, what was most pleasant, I was several times plied with real Irish Guinness.

Next day the idea of testing was realized. The Brazilians brought a woman who agreed to play the guinea-pig, and all of us took and analyzed her pictures. The process was several minutes long; I was writing results of analysis another half an hour and passed them to the Organizing Committee. The patient was tested right in the lobby of the Congress and processing was projected from computer to the large screen.

Special rooms were required for Rosemary Steel from England and Peter Mandel from Germany: they both used photographic paper and, therefore, there should have been no light. Then the photographs taken were brought to the studio, and the results were ready by the evening.

Newton Milhomens from Brazil uses his own equipment: a hand is placed into a lightproof camera, and a specially selected color film is used. Then this film shall be processed and color photographs shall be printed. Not a quick process, so analysis can be started only in a few hours.

In the morning of the next day the Chairman of the Organizing Committee doctor Roberto Leute opened the envelopes and read out their contents one after another. The main conclusions coincided, which produced happy smiles and applause among the attendants. It became clear that notwithstanding the difference in directions and equipment, we all used the same principles to research a patient's state. Therefore, we can speak about bioelectrography as a diagnostic direction which has different approaches. Our approach involving computer processing is the most modern, but it requires understanding of modern techniques. For people of older generations photography seems to be easier and more accessible, and primary equipment is significantly cheaper. However, then you have to constantly spend money on photomaterials, which often becomes a tidy sum. And, of course, a special question is storing of results. We can store practically any data sizes on computer carriers, photomaterials do not give this opportunity. In Doctor Mandel's center in Bruhsal, Germany, several rooms are filled with shelves of Kirlian photos of patients. Go and try to find something in there! And what if you have to change your apartment? Or, God forbid, a fire breaks outs? That will make one think...

Brazilian ritual dancing

OUT OF BODY EXPERIENCE AND OTHER TRANSFORMATIONS OF AURA

The light of Wakan Tanka is upon my people
Kablaya

In the middle of 1960's a book by Robert Monroe "Journeys out of the Body" was published in the USA. It described feelings of the author whose spirit separated from the body and began to travel in space independently. At first it scared him very much and he tried to come back into his body. Gradually, as he made sure that nothing fearful went on, he started making his spiritual travels longer and not just looking at his body at rest from the outside, but flying away from it quite far. As mentioned by Monroe, he could fly through the walls, visit people's houses, known and unknown places, in short, take up good tourism without any inconveniences for the physical body and bank account. Monroe even developed a technique of learning to travel outside of their physical body for everyone who wishes.

The book became well known in the West, there appeared researchers, scientists determined that the process of leaving the body is similar to autohypnosis, and the brain really passes into some special state, which is different from both wakeful state and sleep.

Robert Monroe was a talented person. He had an enterprising spirit typical of an American. He managed to create a school featuring his technique, started a journal and published audio cassettes and books. As he began to be successful he eventually created his own institute. The important achievement of this institute was that they discovered the fact that certain low-frequency sound signals given through earphones stimulate transition into the state of spiritual travel. Monroe passed away in 1987, however, his institute continues to work successfully. So, being in Virginia, I agreed to Bob Van de Castle's offer to visit this institution.

After breakfast we got into the car and drove through the forest expanses of Virginia. The first item on our agenda was a visit to the Buddhist Center. The prior of this Center, lama Tenzin, had been present at my lecture and had invited us to visit his estates. After an hour travel we left the main road and turned onto a dirt road. At once I recalled our bad motorways in Russia, as even the dirty roads were well made in Virginia, without deep holes and ruts. We used to say that there are no motorways in Russia, only directions. We crossed the river and, having gone up the hill, entered the gates of the Buddhist Center. Parking proved to be quite good there.

The Center occupied a wonderful place: on top of the hill, surrounded by wooded slopes with branchy trees. Virginia is situated in the same latitude as Turkey, and it is quite hot there in summer. The original occupation of Virginian gentlemen was growing tobacco, and until now, notwithstanding all American anti-smoking campaign, you can see tobacco plantations in some places. All over the country there are a lot of forests, and American fallow-deer bravely come to the road, and squirrels sincerely think that the human being is a kind of source of nuts.

Lama Tenzin was very young, about 30 years by appearance, but it is very difficult for us to determine the age of the Tibetans. Most of his life he had spent in India, had got good education, and 12 years ago had been sent to the USA by the Dalai Lama, in order to preach Tibetan Buddhism.

What a paradoxical situation: the Chinese had captured Tibet, had destroyed hundreds of monasteries, thousands of monks. The Dalai Lama had had to escape to India. A horrible harm had been done against peaceful, friendly people. The world had condemned this deed, but nothing had been done. Even in our days people with slogans "Freedom to Tibet" walk around the Trafalgar Column in London - but what's the sense? And as a result of those political processes, the once fully closed and secret schools of Tibetan Buddhism have begun to spread in the world. Tibetan sacred books were translated into European languages, their interpretations and liberal translations appeared. It has turned out that Buddhist philosophy with its peaceful acceptance of good and evil, fundamental attitude to death, inner spiritual essence of human being as the basis, has greatly contributed to teaching the soul of Western people. European interpreters of sacred texts have appeared, but they haven't managed to always penetrate into the core of the ancient teachings themselves. Therefore, the Dalai Lama constantly sends his emissaries to the world and spends a lot of time outside India himself. An interesting question: if the Chinese had not captured Tibet, would this secret religion have become widespread in the world, or would it have remained behind the ice barriers of the mountains? What is Good and Evil in this world? Isn't Devil the antithesis of God, created by God not to let God to become too proud for the results of his work? Evil makes a step, so that God could come forward and show itself. There is no place for development in perfect world with absolute balance.

Tenzin managed to justify the hopes of the Dalai Lama. He created centers in USA, Mexico, Poland, and Russia. He wrote a number of good books and published them in different languages, including Russian. Hundreds of people gather at his Center in the Virginian hills for workshops, festivals, and holidays. They come together with their families, children, friends, and many of them find a cozy spiritual place there. You don't have to say you are a Buddhist to come there. People of all races, nations, and religions are accepted in the Center.

Tenzin guided us around the territory of the Center, showed halls for meetings and meditations, educational center, and the hotel under construction. After that we had a nice vegeterian lunch and pleasant discussion on the open veranda. The food was tasty, healthy, from natural plants and vegetables, and this was a pleasant exception in our American ration. (The fact is that from a European point of view usual American food is tasteless, synthetic, and detrimental to health. This nation suffers from many diseases, first of all, because of the character of their usual nutrition. But the topic of healthy nutrition we will discuss another time).

We left the Center of Tenzin in a most wonderful mood. And it was associated with our energy level, as was registered on the energy field picture. Compare my picture before and after visiting the Center: (Fig. 14,15).

From the Buddhist Center we made our way to the Monroe Institute. It was only about 20 minutes drive, and for this time, sitting near Tenzin, we discussed his next visit to Russia and St. Petersburg.

In the beginning of the Twentieth century an outstanding Buddhist physician Peter Badmaev had founded a Buddhist temple in St. Petersburg. Badmaev had been a wonderful doctor and healer, applying all collection of Oriental medicine very effectively. He had come to St. Petersburg in the middle of 19th century with a small case of instruments and dried herbs and in some few years he had already got many clients, including even members of the tsar's court. This had given him an opportunity to build a large clinic in Saint Petersburg, and luxurious coaches had always been standing at its entrance, however in the days of free reception the poor had also crowded. The Buddhist temple, constructed with imperial permission in the nearest suburb of St. Petersburg, had become the center of attraction for the Buddhists from all North-West of Russia. The Bolsheviks had not been that tolerant as the monarchy, and so they had turned the praying people out of the temple, had made it a warehouse, but it had been so well constructed that even 70 years of desolation had not destroyed it. Now the temple is given back to the Buddhist community and in 15 minutes from the city center, near the beautiful park, one can enter the temple and listen to the plangent singing of the monks.

While we were talking with Tenzin we reached the Monroe Institute. Good parking. Beautiful complex of buildings. The Americans show good taste in building design, even using plywood. 90% of buildings in the USA are made of bars trimmed with plywood. Quick and cheap. Faced with stones or wood from the outside and looking quite presentable. But, in fact, such houses sway from the wind. And if a hurricane comes, it can easily take such a house and carry it to the unknown lands. As it was with the girl Dorothy in the tale "The Wizard of OZ" by Baum.

We were met and accompanied to the laboratory. Robert Monroe had been able to make his techniques widely well-known, introduce them to Pentagon and secret cervices, and receive recognition. For many years the Monroe Institute offers everybody a week course of personality transformation on the basis of special procedure. Training of this type was offered to Tenzin and me.

We took the initial pictures of fingertips. Then I was guided to a small room padded with black material inside. Practically all the room was occupied by a water bed, and when I lay on it, my body was suspended in a state of semi-weightlessness. Big earphones were put on my ears, the door was closed, and I found myself in total darkness. A thin melody sounded in the ears, and soon it filled all my being, all the space, breaking over the body as waves. The tune of the melody was changing from time to time, but it was not even a melody, it was the sound of waves, crackling of the silence, rustling of the darkness. Sound waves were pricking the body, stretching or pressing it, it was a peculiar sensation, although not the most interesting I had ever experienced in my life.

From time to time a soft voice inquired about my state, though there were no devices measuring the state. In about 10 minutes the session ended, and colleagues helped me to come out of the chamber. They certainly helped me because I was stoned and my head was swimming. That was clearly seen in the picture of my field: (Fig. 16).

Nothing remained from the good state after the Buddhist Center.

This was the goal of the session, as explained by the staff of the Monroe Institute. Ultra-low frequency sounds were applied through the ear-phones, and frequencies in right and left ears slightly differed. This caused desynchronization of the work of the right and left cerebral hemispheres. (Remember that we already discussed problems of the right and left brain). A human being left his usual stable state, and common perception of the world was violated. Many people suddenly understood that the outer world could be perceived otherwise, not just in its usual way, and this evoked the understanding that it was possible to live in different ways, and take everything which happens in one's life from different points of view. I heard from some Americans that a week course at the Monroe Institute helped them to fully change their attitude to life. To come out of the routine flow of life and problems and make the next step as a new person. Our life mainly depends not on the fact where we live, but how we think about it. Perhaps, the Monroe technique of the frequency influence might turn out to be useful for the psychological rehabilitation of veterans of war, refugees, people who survived hard psychological traumas. As well as for many people suffering from neurosis, heightened irritation, and inner dissatisfaction. To push from the usual platform, to break the routine circle, to get away from numbness, in order to then sit and think: what shall be done next?

I started doing analysis of the dynamic change of the field. (We took time dynamics from one finger within several seconds, and then carried out Fourier-analysis of spectrum). As a result, we obtained the curve (Fig. 17)

characterizing the main frequencies of the energy field. Based on this curve, the main frequencies were calculated (Fig. 18). Computer calculations took a couple of minutes, and when I demonstrated the results, the Institute personnel started to examine the curves with special attention. Then one of them said:

"But these are the same frequencies we applied to your ear-phones. And this graph (Fig. 18) is the frequency curve which appears when two frequencies shifted relative to one another are superimposed. It took us two years of work to obtain such a graph from the analysis of encephalograms. And you did it in 5 minutes! Astonishing! How did you manage to do that?"

After that we were discussing for half an hour what all that would have meant and what we should do with this. Both parties were satisfied. They, because their ideas were proven, we, since we managed to do that. We said good-bye to each other in the best mood, although it hardly influenced the picture of my field: my head was swimming. I didn't know whether I was coming or going.

Skeptic. Your examples with the change of energy field are not very convincing. All this is self-suggestion. They inculcated you with their ideas, and you fell under their influence.

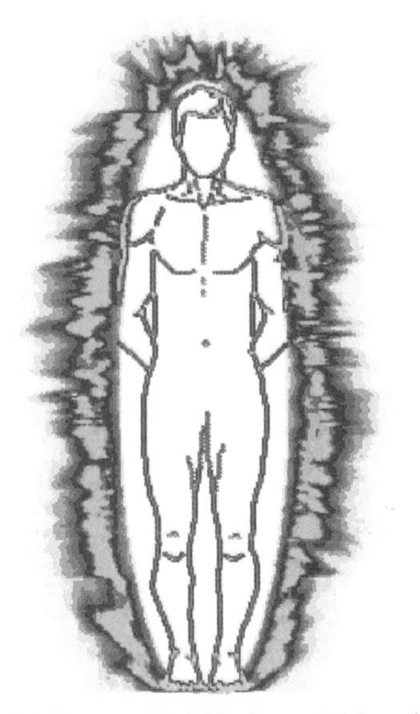

Fig.14. Energy Field before visiting the Buddhist Center.

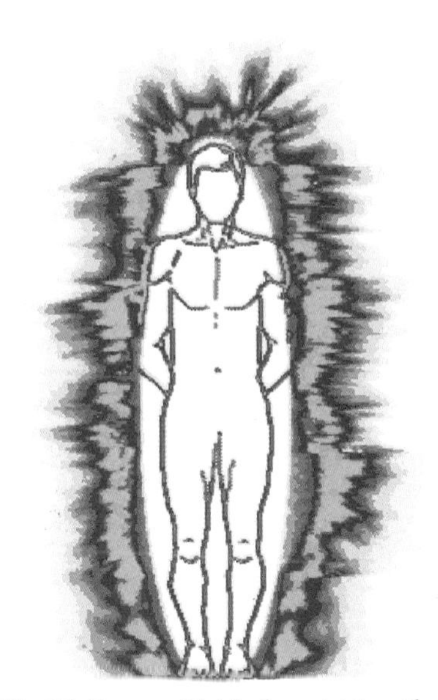

Fig.15. Energy Field after visiting the Buddhist Center.

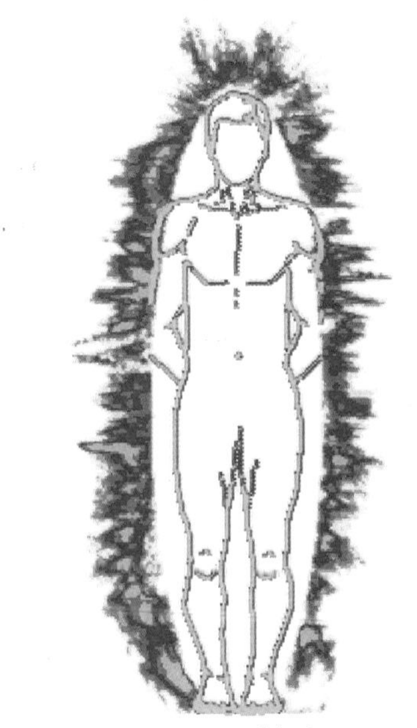

Fig.16. Energy Field after visiting the Monroe Institute.

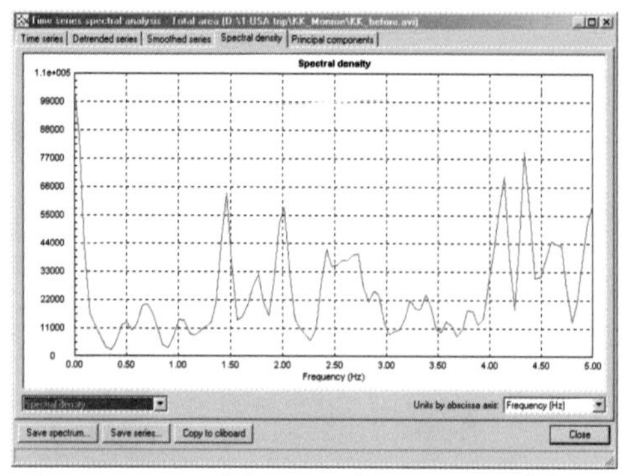

Fig.17. Fourier spectrum of the main frequencies of the energy field.

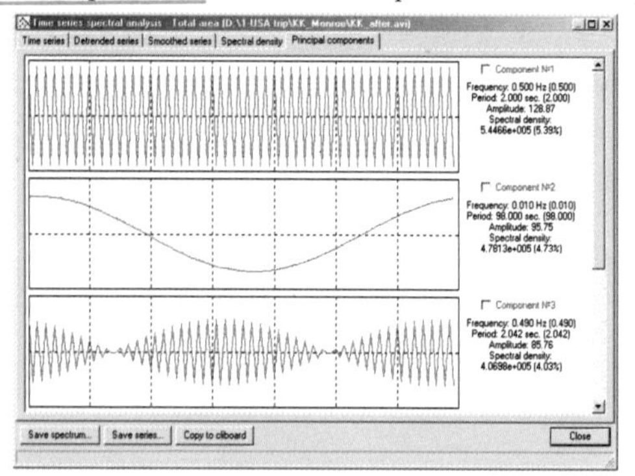

Fig.18. The main frequencies of the energy field.

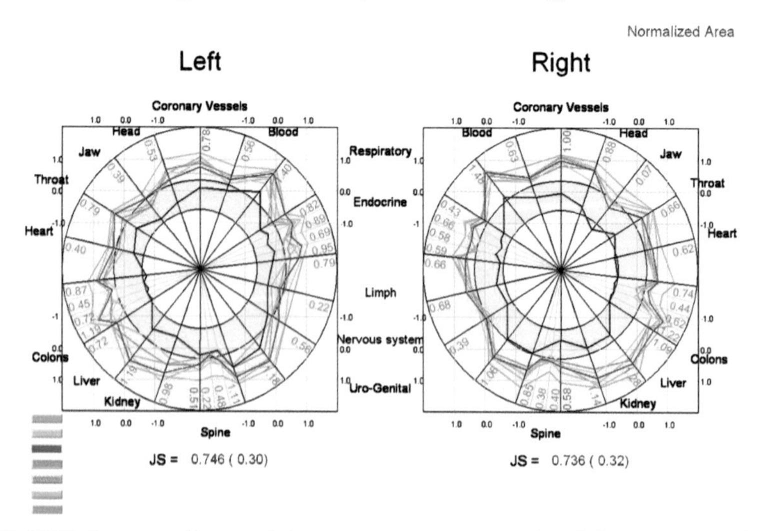

Fig.19. EPC diagrams of one and the same person, measured in different points of time during one year

Author. First of all, the result is important in many cases, even if we do not fully understand what is going on. If the head aches, but eases off from a finger flick, let us use this flick as a method of treatment. Then some day scientists will investigate the deep mechanisms. Of course, if there is somebody interested to investigate. Second, self-suggestion is an important component of treatment. If you wish to be healthy - be healthy. Pessimists get ill more seriously and more often. And, finally, there are methods of objective estimation of state, and bioelectrography is a very sensitive method, but it is not so easy to affect me. So, if there is influence, you can not hide from it.

Skeptic. All your changes might be just variations. Measured now - one result, in an hour - absolutely another. Body temperature, for example, can show you a lot: 36,6 - good, 37,2 - bad. Your pictures are something different.

Author. Finally I heard a serious question from you. Your professional skill grows. I can give you quite a substantiated answer. A special research was performed with a group of ill and healthy people. They were measured with 1 hour and 1 day's interval. 80% of participants showed that the variability did not exceed 6-8%. For some people these numbers were higher, but only for the EPC-images taken without filter, i.e. for psychological field. Parameters remain invariable within many months for people with stable psyche and normal state of health. Deviations indicate coming diseases, potential problems. For instance, look at the EPC diagrams of state of one and the same person, measured in different points of time during one year: (Илл. 20)

As we see, 8 curves quite exactly reproduce one another. This means that the state of the person remained very stable. And now pay attention to the curve in the center. That day the person poisoned himself with bad sausage, and his energy level had drastically fallen, particularly in the area of intestines. What was more, the measurement had been done before he felt the symptoms of poisoning.

Skeptic. Are you saying that you can foretell diseases?

Author. To some extent. To put it more precisely, we analyze the places of functional energy deficiency, which can cause problems in the future.

Skeptic. And so, you had cases when your predictions were proved?

Author. Unfortunately, yes. It happened several times in my practice. People do not always pay attention to warnings. Here, one example. Once visiting a big company in New York I told the Vice-President that he had a potentially weak heart, and it was functionally overloaded. He replied that he had never had problems with his heart. With this we said good-bye to each other, and I left without a contract, at that. In half a year I was invited to this company again, and again I met the Vice-President. He shook my hand with respect and told that a month after our first meeting he was taken to hospital with a serious myocardial infarct and had spent several weeks in the hospital bed. After this he called his staff and asked to discuss the terms of a contract that would have satisfied me. That time I got the contract signed very quickly.

Skeptic. Here your essence is expressed: to derive benefit even from people's problems. Wait a little; I will think of how to plague you more!

INDEX OF HEALTH

In the effulgent lotus of the heart dwells Brahman the Light of lights
Upanishad

I have a friend, a professor of our University. He is a very energetic person, hale and hearty, and self-confident. He can tell about his projects, problems, and beloved women for hours, and the interlocutor needs to just show minimum attention: he has his own opinion about everything. From time to time Arkady (that's how I call him; students, respectively, Arkady Valentinovich) comes to our laboratory to drink tea and speak about his plans. And to admire his own perfect and bright field. It usually looks as a model (Fig. 20)

But once he came, and the field was much weaker: (Fig. 21)

Arkady was surprised and advised to check our equipment. In order to test the instruments we usually take images of a special titanium cylinder and use this data to determine the stability of work. Everything was in order. Next day the field was even worse (Fig. 22)

Arkady felt very well, so for the whole hour with the wit typical of our intellectuals he was describing what benefit we could have brought to our country and all of Humanity if we had given up our field and had started doing something useful, for example, growing spring onions.

The next day he got a serious flu and spent a week in bed with high temperature.

Such situations are not rare. All potential problems first of all show up in the field picture. These can be temporary functional disorders, which pass quickly and don't leave vestiges, or more constant disorders, which arise as a result of permanent negative factors: bad ecology, wrong nutrition, emotional experiences, stress, or other people's negative fields. The human energy field becomes disrupted, defective, which opens up additional opportunities for harmful agents to penetrate. And in some moment a functionally weak and overloaded system doesn't stand, and the problem passes to the physical level. The person gets ill - symptoms appear.

It is particularly apparent with children. Children's field is very active and unsteady, same as children's mood: if a child disconsolately cries, amuse him - and in a minute he starts laughing merrily. Thus the field, especially psychological, changes several times during the day, reflecting intensive character of information exchange of the child with the environment. At the beginning this fact put us at a nonplus. How should we work with children? How should we make their analysis? As far as we accumulated experience together with pediatricians it became clear that the picture which we observed on the computer screen was the reflection of the real psychophysiological situation. Indeed, the child's field changes, but this is the reflection of real processes. And if some special marks like such "double rings" around fingers appear in the field picture (Fig. 23) this indicates the altered state of psyche, caused by some specific conditions, and in many cases, unfortunately, by drugs.

At the same time, it was found that the picture of a healthy child with filter should fully meet general rules: i.e. be even and bright. Deviations indicate the present or potential problems with health. Or, in other words, functional disorders. This picture is more stable for adults, for children it changes with the change of functional state, but all in all it doesn't lose its diagnostic meaning.

Sasha was a normal, well developed 5-year-old child. He attended kindergarten, sometimes had childhood diseases: fevers, scarlatina, cold in the head. The only thing which worried his parents was his periodic state when Sasha would lie down and say that he had "no power to move". He lay for half an hour, and then ran to play again. Parents brought Sasha to doctors, made different analysis - everything was more or less normal. Then they came to take a picture of his field with filter - it was all disrupted and weak. It was unclear which particular systems are connected with this. Some general system disorder. Pictures of parents' field were taken: (Fig. 24, 25)

Father had dense and closed field, as a turtle's shell, mother's field was powerful and super active. Our psychologist spoke with the parents. Father - a typical engineer, worked in a state organization, with a lot of responsibilities, very low salary put into his card, so as not to feel ashamed of one another coming for salary. The mother - a new Russian (nouveaux riche) business woman, who managed a publishing house, and her husband and son. These people always have a special type of field: they spread it in the space as tentacles, actively influencing all surrounding people. Naturally, no time for family life and playing with the child.

We advised to somehow change the situation: to change the way they treated the child and to revise his nutrition. It would be cheaper to hire a nurse, or father had to quit work. The mother listened silently, thanked and left. Nothing really was done. In two months Sasha's blood formula significantly changed, though before that his analyses had been quite good. He was taken to hospital and treated, but in our opinion the reason of the disease was not on the physical, but on the psychological level.

Such impressions led us with Prof. Pavel Bundzen to the creation of "health index." -index, which would characterize the human state as temperature and blood pressure. 36.60C (980F) - everything is OK, 370C (990F) - some flaccid inflammatory process in the organism, 380C (1000F) - already a disease. The Biofield's picture has the same peculiarities. Individual features are present, but there is something common. How to describe this? After long experiments and discussions with Prof. Bundzen we came to the conclusion that such index is the integral coefficient, calculated in the "EPC Diagram" program: (Fig. 26)

I will not bother you with the description of how this coefficient is calculated. You can find this in our publications . In order to test our idea we asked all our colleagues to measure healthy people and send us the results. Many people replied and in a month we had a database of practically healthy people from Sweden, Germany, USA, and, certainly, Russia.

Skeptic. And how did you know that they were healthy? Did you have results of blood, urine analyses, clinical trials?

Author. No, we did not have that in most of the cases. Clinical trials require several days, a group of specialists and equipment. Therefore, we are speaking about "practically healthy" people, i.e. people who had no serious diseases or, as usual, chronic problems compensated at the expense of other physiological systems. As soon as the data was gathered, statistics began to work: the calculated integral coefficients formed certain domain for the majority of people. This domain was taken as normal based on the laws of statistics. What was more, the size of this domain depended on the age. The ratio between the coefficients of right and left hands characterized balance of brain hemispheres.

Skeptic. And where did you take this from?

Author. This is one of the results of our long-term research, and we will discuss it next time.

Skeptic. What for do you need this index? Where can you use it?

Author. Once we participated in the examination of people from the Chernobyl zone. A first problem which arose for medical commissions who came to carry out examination: what specialists should be taken and what analysis should be made? The budget was always limited, a great number of people should be examined, and nobody had any idea about their health. Here the index of health was right to the point. It allowed making selection at the first stage: when the person was healthy or when the state of the organism required attention. In the first case a person should go home, and in the second - the analysis should be made and specialists should be consulted.

Mass medical examination at school, at work, in the army has the same goal. But in the everyday life the index of health plays an important role. An example from my practice.

June and July is a very tight time in our work. Completing and handing over reports, preparing and holding the Congress, a couple of workshops - work 12 hours a day 7 days a week. Therefore, I was not surprised when my EPC images became very bad in the middle of July. "OK, the Congress will end, everybody will leave, I will have a chance to peacefully sit in my summer house, make conclusions, prepare a few articles, and everything will improve", I consoled myself. What was more; it was hard to assess some particular systems and organs: the field became entirely small, the EPC diagram "shrank" and mostly passed into the negative area.

August came, I could have peaceful rest, sitting at the computer, my physical state seemed to be normal, but EPC images did not revert to the norm.

And on August 5 (I can remember it was Saturday) what I felt was like a splinter in a heel. Examined - nothing was seen. Just in case, swabbed with iodine. I was scalded with pain as if with boiling water and temperature rose by Sunday night. Then from bad to worse. Abscess, lymphadenitis, temperature. It resulted in two surgeries, a week in hospital and three weeks of out-patient treatment. It became clear

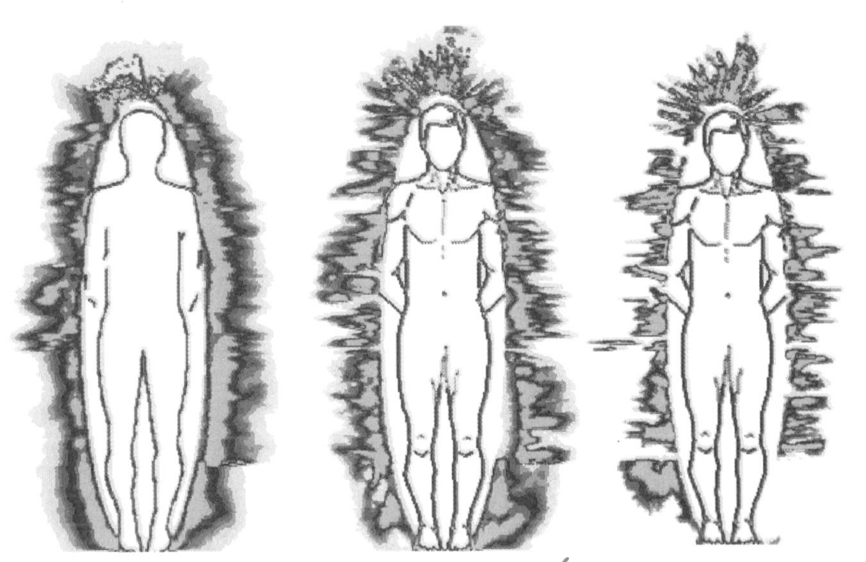

Fig.20. The picture of the energy field of Arkady.

Fig.21. The picture of the energy field of Arkady one day after.

Fig. 22. The picture of the energy field of Arkady two days after.

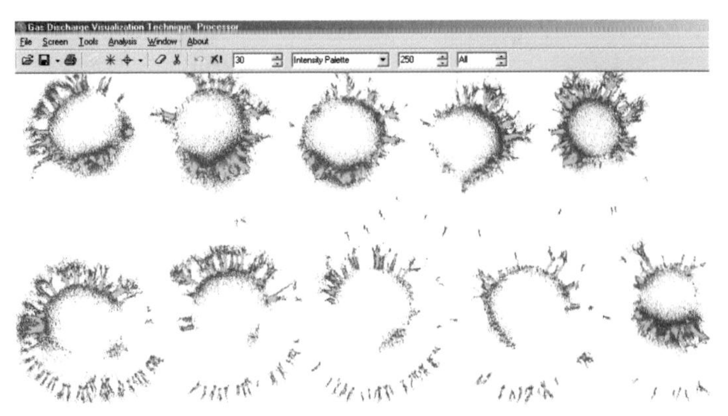

Fig. 23. EPC images of fingers.

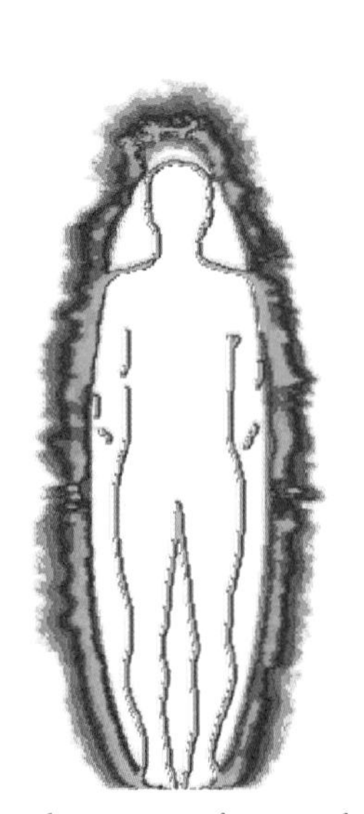

Fig.24. The picture of energy field of father.

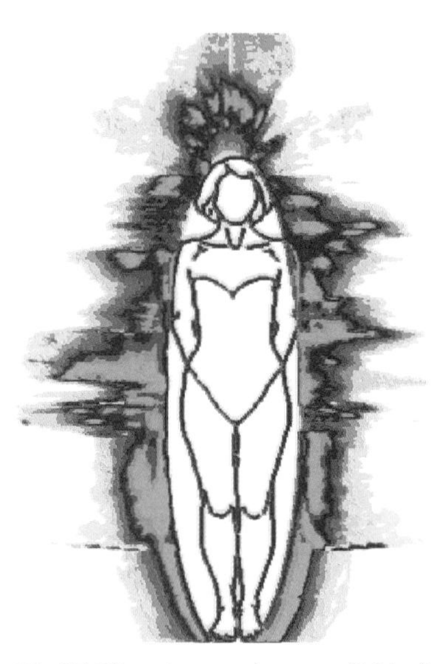

Fig.25. The picture of energy field of mother.

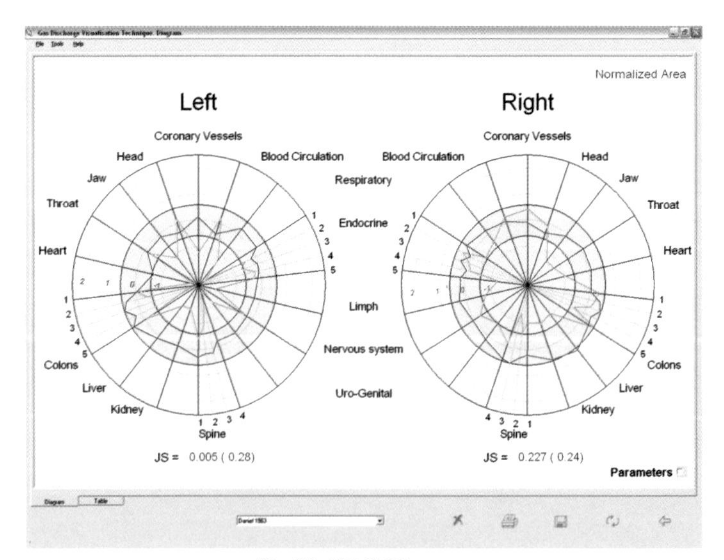

Fig.26. EPC Diagram.

to me how Bazarov from Turgeniev's novel "Fathers and Sons" could have died of cutting a finger. Now we have antibiotics, plasmophoresis, blood photomodification, and what was then? Only bloodletting and a priest.

This entire unpleasant situation was first of all connected with general weakening of the organism, loss of immunity, as well as with the ingress of some foreign agents. Could that have been determined beforehand using standard methods? Theoretically, yes. Practically - no. Identification of immune status by blood is quite a complex and expensive analysis. EPC bioelectrography method enabled to easily and quickly determine that the index of health fell lower than the norm. What can be done in this case? First of all, it's necessary to increase the energy level. Not to look for the cause of energy deficiency, to take the necessary steps right away. The best I could have done was to go somewhere to the mountains for a week, walk mountain tracks and drink glacial water. If this was not possible, to use the system of complex immunocorrection. If it did not help - look for the cause of energy deficiency, not waiting until the disease developed. And another thing, very important - to meditate, pray and try to understand what is wrong in your life, what is your sin in the face of God.

By the way, after treatment my EPC images came back to normal. But still it was necessary to go to the mountains.

Thus, evaluation of the index of health represents a new approach to the control of state. Correction of the state of an individual, not treatment of the diseases. Let us avoid bringing ourselves to disease. Let us support our body and soul in active working state.

"Who needs this?" you will ask. Each of us! In order to be healthy, cheerful and full of energy, not to get into hospital breaking all life plans and then to earn money for pills during half of your life, not to loose life tonus and correct brewing problems when convenient for you.

I believe that in some time systems of monitoring and long term control of health state on computer basis will start being actively developed in the world. Use a device once a day, place your hand, the device takes a set of parameters and sends to the Internet database. One can always take this device along. Parameters are in the normal range - everything is OK. Beyond the normal limits - and nice lady's voice warns you: "Mr. Johnson, please, pay attention to your liver. Refrain from the third glass of wine". Parameters are in the dangerous zone - a signal to the doctor: detailed analyses should be made. Thus a disease can be caught at the very early stage, and this makes doctor's work considerably easier: modern medicine effectively copes with the early stage diseases, perfect when no clinical symptoms.

Our colleagues from the Ukrainian city Dnepropetrovsk told about two similar cases with different consequences. In both cases the field picture indicated strong blocks in the head area. Both persons were healthy active men about 50 years old. One of them listened to the advice of doctors, visited massage, started doing exercises and taking natural measures. The picture improved in a month, and then it was entirely in the norm zone. The second client did not believe gave up on the doctors: "Cooking up here". In a month he had a stroke, was hospitalized and partially lost his brain activity. The main principle of preventive medicine - to find out potentially dangerous zones of physical symptoms and correct the disorders using the most natural methods. Of course, we are doing only first steps on the way of preventive medicine, and many questions require understanding. For example, the following case happened not long ago. A man of about 40 years old. We measured the field - a low index of health. Nothing particular was found. We advised to take a rest and eat vegetables and fruit. In a few days he pricked a finger with a bone. Next day - temperature 39°C and a very serious inflammation. A month in the hospital. We are speaking about weakening of immunity, but in the majority of cases these are just words. Bioelectrography technique enables to quickly and easily evaluate the index of health, field picture, and, thus, estimate the general state. This data can be transferred to the central database by Internet, where it will be stored during the whole person's life.

Skeptic. And what about personal data secret? Each hacker will be able to get into the Internet and download your data. And then a person will not be employed because his field is crooked. This smells of total surveillance.

Author. As if now it is impossible to go to the hospital and take the health history of any person! A little bit of persistency - and you got it. But on the other hand, you are right: information society requires more and more openness from every person. Individual number, tax history, credit history, insurance history, car insurance. There are a lot of advantages in this, although certain inconveniences. Western society becomes open. No choice but to hope for the development of improved collective mind. Later on we will discuss this.

AFRICAN ZOMBIE AND WORLD TERRORISM

The Lord is my light
Whom shall I fear?
Psalms

For many years one of the main topics of our research has been the Altered States of Consciousness (ASC) 4. What is this? Imagine that you live an ordinary life: you are occupied with everyday activity doing your job, meeting other people and thinking where to eat something tasty this night. If we ask you by the end of the day where you have been and what have you been doing you will more or less clearly describe your activity. But let us consider another situation: a scientist writes a treatise of his life, a student gets ready for the state exams, a youth in love imagines the eyes of his beloved girl. They are so preoccupied that they do not notice anything happening around. Another striking case: a sleepwalker who can walk along the roof ridge and never loose balance if nobody wakes him up. If somebody does, he will fall down. Here is the main obvious difference of the normal state of consciousness from the altered state: in the normal state an individual controls his surrounding, in the altered state one fully lives in his or her own inner world...

At midnight we gathered in a small glade, surrounded by a wall of tropical trees. South America, Peru, February, middle of the equatorial forest. We are sitting silently, waiting, strange presentiment as if something settled in the air. A priest with a drum appears, starts a slow rhythm. It reminds of something? Sure, it is modern rap, but slow. The second priest takes out a clay bottle in the form of a jaguar, pulls out the wooden plug, and pours a little bit of green liquid in a clay glass. He stretches it out to the closest one in the group: "Ready? Haven't changed your mind?" Hesitating for a second, the person takes the glass and drinks the liquid, and with evident relief lies on the mat. The priest holds out the glass to the next one. My turn comes. There is no way back, so I resolutely take the glass and drink it up. A grassy bitterish taste. No alcohol. Now we have to wait until it starts its action.

The course of this ceremony called Ajurvasco is very typical of such occasions. There should be a group of people, rhythmic music and some drug. Ajurvasco in Peru, Peyotl in Mexico, Cashassa in Brazil, medovuha (honey beverage) in ancient Russian ceremonies.

In half an hour the liquid begins to have its effect. The surrounding landscape distorts, tree branches take a winding shape, they move, as if alive, big villous stars are seen through them. I look around and suddenly see that the glade is full of snakes. They are everywhere: in the grass, under the snags, on tree branches. I am not fond of catching vipers and grass-snakes, but this picture does not cause any unpleasant feelings: the snakes seem to be part of the environment and do not pose any threat.

At the same time, looking around, I suddenly understand where the ancient art of the South American civilizations came from: it is obvious that ancient painters saw the world through the prism of Ajurvasco and Peyotl, and the surrounding in a phantasmagoric distorted form.

The green liquid was doing its bit: everything was boiling up inside and my body was more and more seized with weakness. I glanced around: my companions took most scenic poses: somebody was walking to and fro, squatting down, somebody was running, moving in a circle on all fours, and another one was roaring in fits of laughter. Suddenly my consciousness went blank and I felt I was a jaguar, a powerful cat, I felt strong legs, fangs, velvety fur all over the body, and a supple docile tail. I felt serene and powerful, my bulging eyes devoured the faintest movements in the foliage of the jungles, and my ears were moving slightly gathering all sounds of the night forest. This was a calm of a compressed spring, keeping its energy up to the final jump. A threat - the spring straightens up and the stored energy falls onto the victim in a terrible deadly blow.

That was not the only metamorphosis I experienced that night. The body was squirming in cramps of weakness and sickness, and the spirit traveled in the astral realm.

During the whole ceremony a few people watched the group. They cleaned the sweat from the wet bodies, drew aside and washed away the signs of deep physical cleanings; they knew no shame and no fastidiousness. Everything was natural in that illusive night world. And what was more they took EPC-grams of all the participants with portable equipment from time to time.

See the pictures. First one is the state of a human field before the beginning of the ceremony: (Fig. 27)

Fig.27. The picture of energy field before the Ajurvasco ceremony.

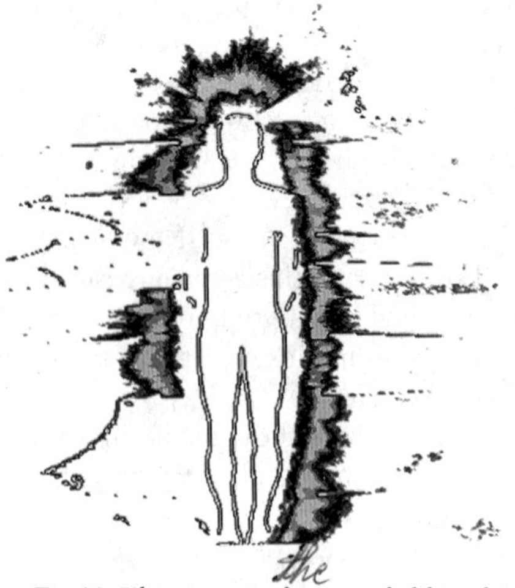

Fig.28. The picture of energy field in the process of Ajurvasco ceremony.

A powerful, strong field, physically and mentally active person. Now another picture, in the middle of the ceremony, when the poison started its destructive effect: (Fig. 28)

As you see, the entire right part of the glow is practically missing and the left one is even brighter as compared with the initial state. But the right part corresponds to the left brain, pertaining to logic, speech, control of behavior. It becomes clear why the participants of the ceremony were laughing, crying, and crawling as children: they could not control their behavior. This is why they were dragged aside periodically. At the same time the left part - the right brain - is super-active. This is the source of emotions, imagination, feelings, and subconscious sensations. All those vivid images and visions came from there. By the end of the ceremony (Fig. 29) the right part of the glow gradually recovered, but this required several hours. Ancient Peruvian traditions say that by the end of the ceremony the field should be "closed", otherwise a person becomes affected by all outer influences and attacks.

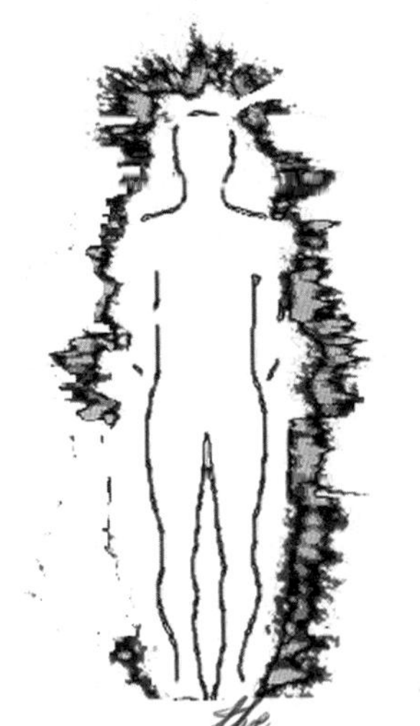

Fig.29. The picture of energy field after the Ajurvasco ceremony.

Our practical experience also brought us to an empirical conclusion that the "breaks" and "holes" in the field picture make the human field vulnerable to the physical and psychological influences. As if, indeed, foreign agents penetrate into the body through these breaks and do their destructive job. Which agents? Bacteria, viruses and protozoa on the physical level, bad thoughts on the mental, and collective hatred on the spiritual level.

Now look at the break in the head area. It is marked with an arrow in the picture. Isn't it like a channel? It remains after the end of the ceremony. Maybe this is really a channel connecting a person with another informational space? American science fiction films play up this topic very often; science makes only first uncertain assumptions.

Sceptik. So, why did you do this? Poison yourself with a strange potion to grow a tail! Go to the Zoo, stay in front of a cage - and imagine you are a beast! A waste of time again!

Author. Such a habit - to try all experiments myself. In order to adequately evaluate stories from people who were tested. And from the practical point of view all this has direct connection to our life. And, unfortunately, a tragic one. I will tell you another story.

In the Middle Ages a fortress rose in an oasis of the Arabian desert. The sheikh, tired of the feats of arms, secluded himself in his lost nook of this world. But the caravan leaders told dark stories about this mysterious fortress and preferred not to stay overnight in that oasis. And the dark stories weren't fairy tales.

The sheikh recruited most skilful and strong warriors from all countries, seducing them with lavish payment. The warriors were training in military art in the fortress and competed with one another. They were praying mornings and evenings, and skilled mullahs taught them surahs of the Koran. The warriors were forbidden to leave the fortress, but they were fed well and the fees were paid regularly. The only thing which was inconvenient for them was the absence of any women in the fortress. In a few weeks of such life the blood started boiling up and sparks of rage appeared in their eyes.

And then the sheikh invited a young warrior for a talk. He tenderly asked about his life, native land, and poured more foamy wine in the bowl. The youth did not notice how he fell asleep, but having opened his eyes he found himself in a beautiful marble hall with fountains and pools, where the blue water streamed and its drops sprinkled the meadows of bright live flowers. But brighter than all paints and more vivid than all the waterfalls were the young beautiful girls, surrounding the youth's bed. They were tender, caressing and accessible, it seemed that everything invented by Humanity in the art of passion and love was accumulated in this hall. Thus hours and days had passed, in passion and feasts, but suddenly, having awaken, the youth saw the known stone loopholes of the fortress, and the sheikh sitting in the same pose in front of him.

"Surprised?" the sheikh asked. "You've visited Paradise, my young friend, and met beautiful Gurias there. Weren't they gorgeous? I transferred you there with my magic, and a moment in this life seemed a week in that one. This entire castle is built with my magic. Do you want to serve me, so that I will show you the way to Paradise, or you will prefer a usual boring mode of living?"

Usually the youth agreed unhesitatingly, and those who didn't, did not live long. And the warrior was taught the art of hiding and killing, not with the weapon, but with any other improvised means: from pin to finger. And little by little a thought that if they die by the order of the sheikh they would go to this Paradise, to the Gurias, was put into the mind of all these warriors.

A few decades later the warriors of the mysterious castle, Assassins as they were called in the East, started playing a significant role in the policy of Arabic countries. The sheikh sent an ultimatum to the ruler of an area or country ending with the words: "If you don't carry out my requirements by the end of the big moon, you will die". The rulers threw away the paper with laughter and looked at the ranks of their faithful guardsmen, but at the proclaimed time a snake crawled out from the sleeve of a dervish and bit the ruler, or a visiting merchant, bowing, brought a Damascus saber to the ruler and plunged it immediately into his chest. The guards tore Assassins in pieces on the spot, but they did not even resist as they were anticipating their meeting with the Gurias.

The mystic fortress intimidated the rulers of the East for centuries, until it was wiped off the face of the earth by the troops of Tamerlan, who destroyed both warriors and Gurias with a horse raid.

Doesn't this story remind you of something? Boeing airliner attacks on the New York skyscrapers, Palestinian boys blowing themselves up in the buses and discos of Israel, Japanese kamikaze storming American ships in "one-time" planes, Chechen terrorists taking hostages in a Moscow theater and Beslan school? A person means nothing to terrorists and rulers. Just a pawn in the play.

Skeptic. But how does this relate to the state of consciousness? The lads are paid well and proclaimed heroes - so they are ready to give their lives.

Author. It's not so easy. I will promise ten thousand to your family if you jump from the tenth floor. Don't want to? That's it. Money is not the most significant thing in these situations, although it also plays its role. The most important is a special training of consciousness, which is transferred to a special state with a series of methods. Important elements: training is performed in a group, whose participants strengthen the intention of one another, rhythmic music is used, or rhythmic frequencies, tuning up one's brain and slogans are advanced, so that a person feels his uniqueness, peculiarity, and difference from all other groups. This is a special, highly professional training, based on strict selection and subtle knowledge of psychology and bioenergetics.

Evil forces use most secret mechanisms of human consciousness. They have perfected their devilish techniques and direct them against Humanity. American employees, Israeli students and Russian housewives turn out to be similarly unprotected against treacherous attacks of the kamikaze. When one's own life is not important, who will think about somebody else's life?

Skeptic. And what is the way out of this? What can we do to withstand this Evil attack?

Author. Only one way out - distribution of information and openness. The more open the society becomes, the more youths join the highest achievements of civilization, and the less the conditions to bring up kamikazes can be created. When a person sees perspectives in life, he doesn't want to say good-bye to it. When nothing good is in life, no regret comes about it. And, naturally, it's necessary to investigate consciousness, the powerful force which is capable of making great transformations and great evil.

THE SPIRIT HOVERED ABOVE THE WATERS...

The Light of Divine Amaterasu shines forever
Munetada

In November, 2001, I was invited to speak at the World Parapsychology Congress in Basel, Switzerland. The Congresses in Basel are held every year, already for 20 years, and gather several thousands of people from Europe and all over the world. Each year they are dedicated to different topics related to strange phenomena; this time the topic was healing. We will discuss this question in more details some day, now I would like to describe just one direction of research which was started at this congress and has become wide spread.

Basel is a nice small town near the French border. Even the airport is in France, and a special corridor leads you from the arrival hall to Switzerland. The old city was constructed in the Fourteenth through Sixteenth centuries, and taking a walk along the narrow streets, you feel a real spirit of the Medieval Europe. Crossing the bridge you find yourself in a modern standard city where one of the points of interest is the Congress Hall. A big six-store building, a five-star hotel in one part and forum in the other with several big halls, a great two-level foyer and a presentation floor. A part of the World Parapsychology Congress is an exhibition. Hundreds of companies and individuals offer magic crystals, charged amulets, Indian, American Indian, Shaman objects, oils, aromas, food supplements, books, and cassettes. The Medieval inquisition would have burned all these objects as the means of witchcraft, together with the sellers. Now this is no longer the rule, and if something is burned in the squares now, say, bad books of bad authors, this is a means of advertising campaign done in order to increase the sales. Most important in the modern world is to attract attention to oneself; the means is secondary.

Healers work right there in special cabins, our Moscovite Alexey Nikitin among them. Many years ago at a conference in Sochi he got acquainted with a Swiss married couple, and there at the conference, he cured the husband who had an acute radiculitis attack. They invited Alexey to visit them and asked him to give treatment to their relative. The effect turned out to be so strong that from then on Alexey spent several months a year in Switzerland, healing people and giving lectures on healthy life-style. In spite of his being overweight Alexey rushes about in the hall continuously exchanging bows with friends and broadly smiling to all the others. Cordial smiles are an obligatory attribute of life in Switzerland.

The Congress opens at 10 a.m. and does not calm down until 10 p.m. This is first of all the place of meetings and acquaintances. People from all over the world come here; it's a unique opportunity to speak with one another. After my lecture in the central hall and a three-hour presentation in a small hall accommodating 250 places, people from different countries one after another approached our table in the foyer. There were quite interesting individuals among them. The director of a big biological institute in Germany, a political figure from Chile, the head of a clinic in Italy. Once Alexey introduced a lively man about 60 years old with penetrating black eye,

"Konstantin, pay attention to this gentleman. This is a famous German healer Christos Drossinakis."

The famous German healer was Greek with a typical Mediterranean appearance, quick speech, and lively tenacious look. As soon as we met he offered to organize joint experiments with the mass-media and TV people. "Without publicity there are no prosperity", says a Zulu folk proverb. In a few minutes we were surrounded by a crowd of spectators and TV reporters. Christos brought a middle-aged gentleman, and we measured his energy field: (Fig. 30)

Then Christos stood behind his back, his expression became serious, even dark, a standard smile disappeared, and he started to move his hands at the distance from the patient's back. The session ended in five minutes, and we measured the fingers again. The effect was obvious even for not well-informed person (Fig. 31)

The public applauded; with a contented look, Christos began to give an interview to a pretty blonde.

For the next step we measured his field in the initial state and in the process of healing. The initial field (Fig. 32) was all distorted and not uniform, which, however, corresponded to the overexcited state of Drossinakis during the congress. I had strong doubts that we would be able to find something interesting. Then Christos came into a special state and we took each finger in the moment of maximum activity at his command. A few seconds of computer processing, and an image which considerably differed from those of all the usual people appeared on the screen (Fig. 33)

This was not a common case in our practice. Indeed, as strong as an ox at any rate judging by his energy. We repeated experiments several times during the congress, and every time the result was interesting. At the closing ceremony I demonstrated the data obtained with Drossinakis, and he was smiling with pleasure and shaking hands with all acquaintances. We said good-bye as friends, and Christos promised to come to St. Petersburg in order to hold a detailed experiment. I did not pay much attention to this: what don't people promise in the moments of excitation!

But that time I was wrong. In February 2002 we were holding a workshop on EPC bioelectrography for foreign colleagues, and Drossinakis turned out to be one of the participants. The idea was that during the workshop we would organize a number of experiments.

A very interesting and lively company gathered at the workshop: researchers from the USA, Germany, Ireland, Sweden, Switzerland, England, Israel, and India. The well-known to the reader Robert Van de Castle was among the participants. He told that before the departure he had made a cardiogram, and the doctors had strongly advised him not to travel anywhere.

"Considering you are 76 years and have a bad cardiogram, you need to go to hospital, instead of Russia. Otherwise we can expect the worst. However Bob was a fatalist and an optimist. Therefore, he invested in a good insurance and flew to St. Petersburg, although he did not look very good."

In the mornings we gathered in the computer classroom, to study the mysteries of EPC bioelectrography. In the evenings we went to the philharmonic hall or theater - we wished that the guests should get an idea about our culture. One of the days we went for a walk to the suburb parks of Pushkin. The walk was announced as "Trip to the Russian Winter Forest", and the reality justified its name. It was -25°C (-13°F) that day. The sun peeped out, the trees covered with snow were sparkling in the sun. A short walk in the park came out to be enough: the frozen Europeans, Americans and Indians were happy to enter the warm hall of a local restaurant. There they got the idea of taking a glass of vodka as they had returned from the frost! The first glass was followed by the second, then the third one, after which it was found that the Indians, Americans and Europeans were not that bad in Russian folk dances. In the heat of all this revelry I glanced at Robert and saw that he was not looking well: he turned pale and held on to the heart. I approached Drossinakis and asked him,

"Christos, let us start the process of healing right now."

He agreed straightaway, took Robert aside, offered him a seat, and started making his passes. In half an hour having gained color Bob joined the whole group. In the next days Drossinakis repeated sessions regularly, and, of course, we took EPC-grams. Every day the picture became better and better, and Bob joked more and more often and cheerfully. A natural shortcoming of Drossinakis' sessions was that after his influence the skin had red spots: (Fig. 34)

By the way, I observed such an effect with several other powerful healers. But the biggest impression was when one of them, correcting my aching back, asked,

"Do you mind if I burn your skin slightly?"

"How?"

"I will increase the energy for a deeper penetration into the tissues, but the skin usually burns out at that."

"OK, I like to test everything new on myself".

The healer continued his passes, and I felt light warmth in the back, but when I looked in the mirror later I found a second-degree burn 5 x 7 cm in size. It was closing up for about two weeks, but radiculitis did not show its effect for about two weeks after that.

After coming back to the USA, Robert went to the doctor. The latter made a cardiogram and was amazed,

"Strange. Your cardiogram is like that of a twenty-year old guy. I have observed you for a long time and have never seen such good data."

By the way, before the departure Robert Van de Castle approached me and said,

"Konstantin, I want to share one of my dreams with you. Remember Justine told her dream when I was calling her? Remember that I mentioned some device in her

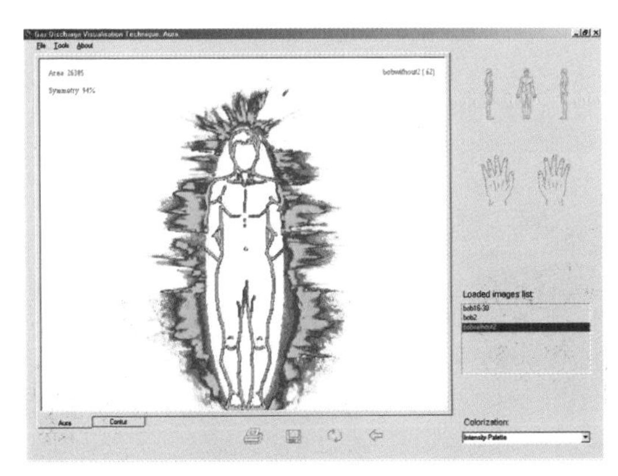

Fig.30. Thc picturc of cncrgy field before the healing session.

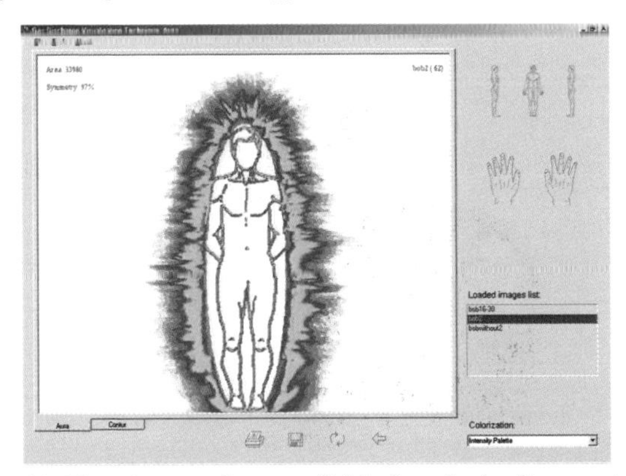

Fig.31. The picture of energy field after the healing session.

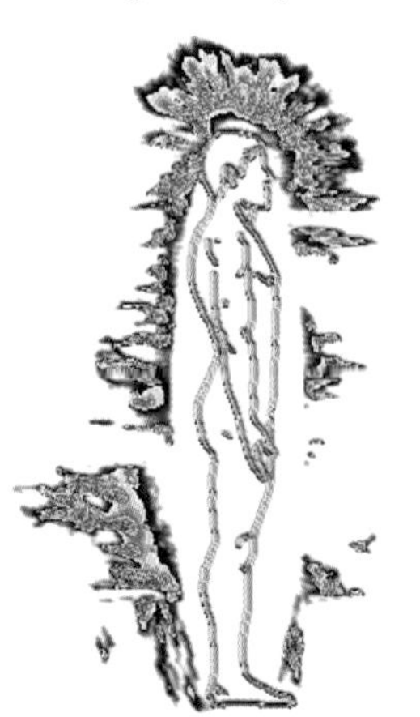

Fig.32. The picture of Drossinakis energy field before the healing session.

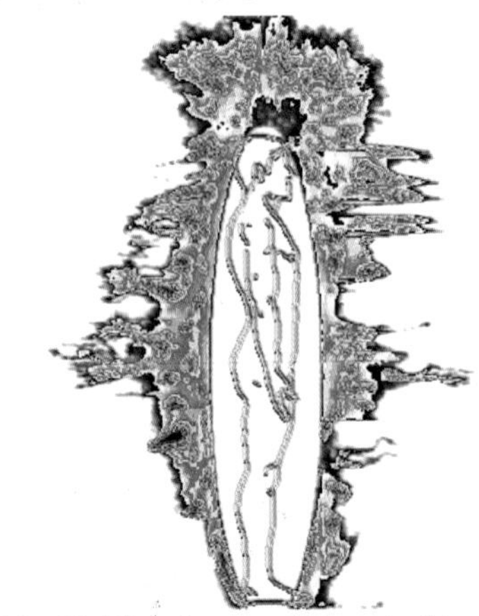

Fig.33. The picture of Drossinakis energy field in the process of the healing session.

Fig.34. Drossinakis in the process of healing.

Fig.35. Measuring mouse.

Fig.36. Measuring mouse.

dream? Well, I saw that device in a dream many years ago. I clearly saw how I was using it in my experiments. And do you know what that device was? What did it measure? Right! It measured energy field of a human being! And now we start a joint program with you! Well, isn't it great? How can one remain being a materialist after such a phenomenon?

I had totally agreed with him. I have seen so many things during the long years of work that it would be enough for another hundred of similar tales. But I still don't stop wondering. Otherwise it gets boring. But let us come back to our winter experiments.

After the experiments with people we passed to animals. I asked a post graduate student, a biologist, to buy a white rat in a pet-shop. We have a special instrument enabling us to measure the energy state of animals. For example, we measure tails of rats and mice, ears of rabbits and horses, noses of cats and dogs (Fig. 35, 36)

We measured the small rat, and Christos started influencing it with his energy. We repeated the measurement - the result was obvious: (Fig. 37, 38)

Next day this experiment demonstrated the effectiveness of Christos' influence again.

By the way, Drossinakis liked that small rat so much that he decided to take it with him to Germany. He smuggled it in the pocket of his suit through all the borders, and when I visited Drossinakis in Frankfurt a couple of months later, I saw there the happiest rat in the world. It walked all over the apartment, slept in the same bed and ate at the same table with Drossinakis, and most probably, by energy state was close to the character of a small dog. Such an upturn in fortune: from a laboratory rat in cold Russia to a domestic pet in Frankfurt.

In the next series of experiments we examined the influence of energy on water. We developed a technique of measuring the induced glow of water, and measured many interesting effects. Water is a magic liquid, the basis of life, it easily takes up information, stores and transfers it. In order to be useful for life - live water, it is hardly sufficient to let it pass through a filter, although this is better than nothing. But this is a separate topic, and we will come back to it. In our winter workshop we were interested in the change of fluorescence of water under the influence of human energy state.

We connected the measuring device to the instrument and began measurement. Drossinakis was sitting near the instrument fixing his stare at the device. In about one minute he said,

"Ready. You can measure it."

We checked the measurement. Indeed, the amplitude of the glow of the liquid sample changed: (Fig. 39)

I wasn't surprised as I expected such a result. And we had done similar experiments with Alan Chumak, Albert Ignatenko, Alexey Nikitin, Victor Philippi, and other skilled native craftsmen. We had even developed a technique of testing intuitive capabilities on this basis. So, I was interested not in the fact of influence, but how big it was. Drossinakis demonstrated quite a strong effect.

The foreign participants were highly impressed with the experiment. For the first time they saw how it was possible to visually demonstrate the influence of a human being on the material world. As water is the basis of any biological system, from an ameba to a human; if you are able to influence the water you can influence anything.

Of course, we repeated the experiment. With a positive result. Naturally, several guests tried to repeat sessions. With no result. No one then had any doubts on the effectiveness of Christos' influence. That fact gave a cause for a few toasts at the gala dinner.

In spring I happened to go to Germany for business, where I was to visit Munich, Dresden, Frankfurt, and fly back from Dusseldorf. In order to freely move, I rented a car. It is a special pleasure to travel in Germany by car, as there are no speed limitations on the highways and it's possible to drive as fast as the car goes. My "Renault" was able to speed up to 190 km/h (110 mph). Only once I could make 210 km/h (130 mph) downhill. You can drive and not be afraid that a local policeman appears from behind the turn.

In Frankfurt I stopped at Drossinakis' small apartment filled with antique furniture. It turned out that in his spare time Drossinakis restored antique furniture, which he gathered at the stores in Greece, Czechia, and France. Christos organized meetings with scientists, journalists; a big article was prepared based on the results of our experiments. While drinking sweet Greek wine he suggested an idea,

"Look, Konstantin, I'm going to Japan for demonstrations in a month. Why don't we influence water from Japan?"

"Great idea, Christos!" I exclaimed, pouring wine into the glass. "We can also influence wine".

"In Russia good wine is expensive, and there's no sense to work with bad one. Come to Greece, I will show you good wine!"

"I'll take you at your word! I will certainly come some day. I have always dreamt to swim in the Aegean Sea. And it's not a bad idea to influence water from Japan."

No sooner said than done. On the agreed day we put 5 similar bottles with drinking water on the window in the laboratory in St. Petersburg. Each had a colored sticker: red, blue, green, etc. At some moment unknown to us, from 12 a.m. to 1 p.m. Drossinakis should have influenced two bottles from Japan, leaving the rest untouched. At 1 p.m. we took samples from all the bottles and made measurements. Naturally, before the experiments, in the morning the samples from all bottles had been taken, which had demonstrated that the water had been the same in all bottles.

Look at the results of measurements. (Fig. 40)

Each curve represents the dynamics of change of the water drop's glow with the course of time, during 10 seconds. As you see, the three upper curves are rather smooth, and the two lower ones demonstrate quite restless behavior: first the glow increases, then decreases, which is typical of active, live water. It is obvious that the character of these two samples differs from all the others. And this can be connected with the mental influence of Drossinakis. The only discrepancy, as it became clear

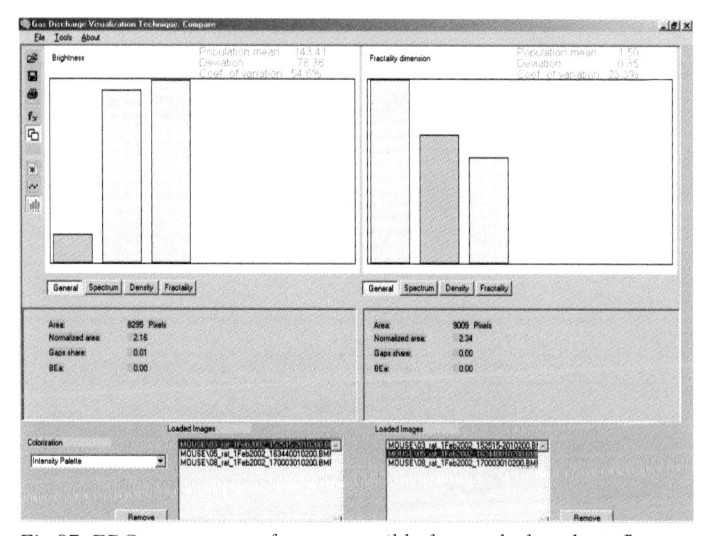

Fig.37. EPC parameters of a mouse tail before and after the influence.

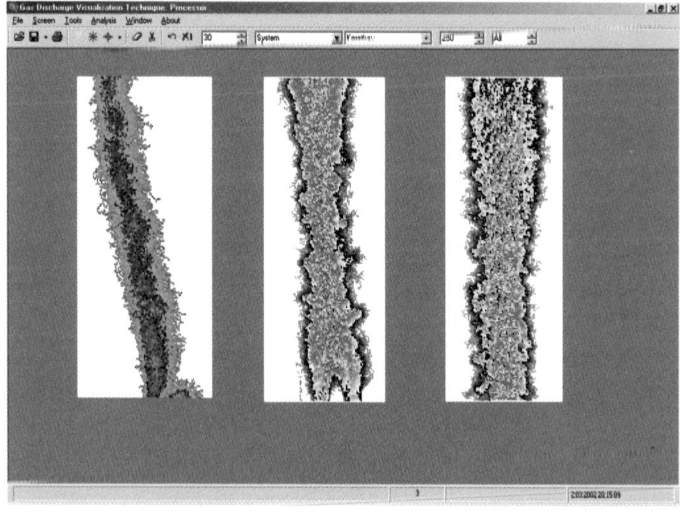

Fig.38. EPC parameters of a mouse tail before and after the influence.

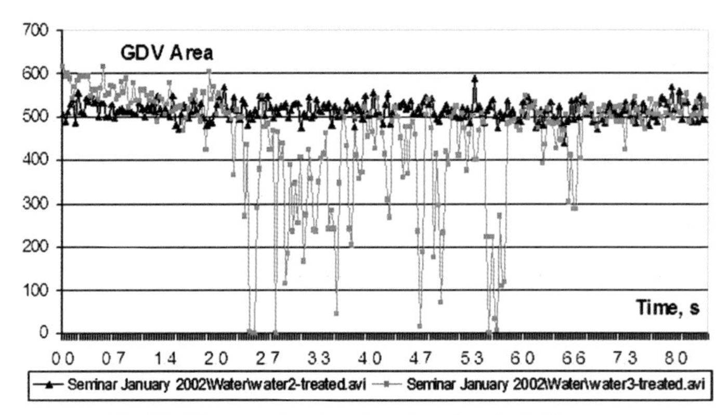

Fig.39. Change of water signal under the influence.

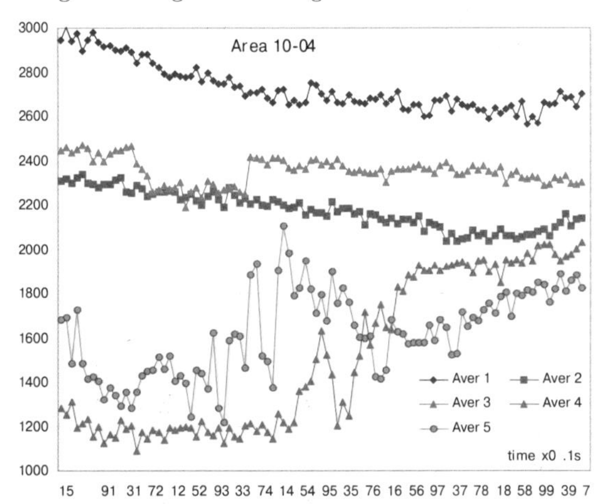

Fig.40. Change of water signal under the influence.

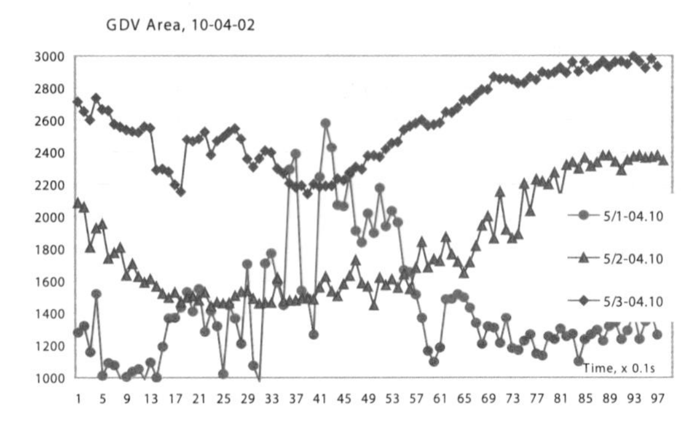

Fig.41. Change of water signal under the influence.

afterwards, was that he had been tried to influence the blue and red bottles, and we observed changes in the green and the red ones.

We repeated the experiment from Japan once again, and twice from Germany. The effect was observed in all cases.

Skeptic. Pack it in! These were random variations. Something got into the bottle. Or somebody grabbed it with dirty hands.

Author. These are the factors which are easily controlled. The staff of a laboratory with good reputation will first of all verify the reproducibility and repeatability of results.

Skeptic. And what happens with your "charged" water then? For how long does it keep the signs of influence?

Author. According to our data, for about 24 hours. It is being transformed, at that. Look at Fig. 41. This is the same sample as in Fig. 40, but measured with a 10 minute interval. As you see, the primary chaotic oscillations are changed with a certain regular curve, which is reproduced within 24 hours. But the character of this curve is still different from that of the initial waters in the graph (Fig. 40). So, some processes do take place in water.

Skeptic. And what processes?

Author. Well, this is a topic of a special discussion. Some other day, now it's time to go and drink tea.

Skeptic. Wait, wait a little. Are you saying that it is possible to just that easily influence the surrounding world with the consciousness?

Author. Well, not that easily, one needs talent and training in order to do that, but the fact exists. All experiments discussed in this chapter indicate that human consciousness can influence physical processes of our world.

Skeptic. And maybe all this is your imagination, experimental errors?

Author. Our results were reproduced in other countries; several laboratories in the world are doing such research, so it is already obvious that these are not experimental errors and not imagination.

Skeptic. This way you will talk to the point that a thought can move mountains, and prayer can raise the dead.

Author. The dead are a special topic, but the influence of prayer is quite measurable. My following story will tell you about that.

CAN A PRAYER BE MEASURED?

Following the light,
The sage takes care of all
Lao Tsu

Once a couple of years ago a nice middle-aged woman approached me at a workshop in Chicago.

"My name is Janet Dunlop. Professor, may I discuss my scientific project with you?" she asked timidly.

"Of course, let us go and sit somewhere and discuss your project", I replied willingly. Several dissertations with the application of our methods and approaches have been done in the USA. Each dissertation is serious research, requiring a few years of painstaking work, and positive results always raise scientific authority of both the author of the dissertation and the methods developed in it.

But when I heard the idea of Janet's work, my first desire was to dissuade her from this. A chance to receive some results seemed to be too improbable. But this elegant, delicate lady turned out to be uncommonly persistent. It seemed that she had been developing this idea for a long time and decided to put it into practice, notwithstanding all difficulties.

Within the two next years Janet attended all my workshops in the USA and Canada. It was astonishing how this lady with an absolutely humanitarian spirit, had made up her mind to master the computer, starting from scratch, and moreover, to use it for a serious scientific work. Dozens of people helped her: one taught her to work with equipment; another one made mathematic calculations, the third one selected data from literature. As a result, Janet made her dissertation, defended it in the University, and demonstrated her level before all the Skeptics.

Indeed, "...when a person wants something very much, the Universe does everything to help". This is from the "Alchemist" by an Argentinean writer Paulo Coelho.

What is the essence of Janet's work? She lives in Missouri - an American remote place. Blocks of nice cottages separated from one another by flower gardens and playgrounds. Each cottage is occupied by one family, and everything on credit - this is more convenient from the viewpoint of paying taxes, even if the family has money. They go to a supermarket once a week; if you forgot something you have to drive another 40 minutes to the closest store or gas station. There are no stalls with beverages there. And all shops close at 7 p.m. If you did not come in time means you came late. People drive to work for one hour or one hour and a half. Every day. As a rule, in the mornings there are traffic jams in one direction, and in the evening - in another. Those who can work at home feel much better; therefore computer home work with Internet data exchange becomes more and more popular in the USA. What is more, there is no necessity to see the boss every day.

In short, American remote places are inhabited with solid people who make their life, having little in common with the noisy gangs and parties of big megapolises.

And in such a remote place, Janet gathered 85 men - volunteers in the age of 35-60 years. Men - in order not to depend on the variability of female hormonal background during several months of experiment. The most active age, although the youthful jumps and fluctuations have already ended. In addition, the men selected should not have had acute or chronic diseases, as well as bad habits like alcoholism and drug addiction. I doubt that it would have been possible to gather such a group in Russia. Moreover, to persuade them to take part in the experiments without a glass of spirits. Conditions of Janet's experiment would have been broken right away.

The state of all volunteers was measured before the beginning of experiments: psychological tests, blood analysis, physiological indices, and parameters of glow of fingers. After that all volunteers were randomly divided into two groups. Nobody of them knew which group he pertained to. And they continued living their usual life: went to work, played with children, and visited church on Sundays. The only difference was that each of the members of one group was prayed for in the mornings and evenings, during a whole month.

The topic is not new for Western science. Relation between science and religion, materialism and spirituality has been a subject of active discussions for more than a century. Elena Blavatskaya from Russia, a genius dilettante, the founder of the theosophical movement, was one of the first who raised this question.

In the end of the Twentieth century this topic changed from a subject of theoretical discussion to an object of scientific experimental research. And the first aim which arose for many believing scientists was to prove that the belief, and particularly the prayer, exerted positive influence on the human state. Indeed, if God helps the believers, he has to give them good health first of all. Not for nothing are there so many stories about the longevity and health of monks. But there this can be explained by the influence of tranquil monastic life. And what about common people, believers and non-believers? Is there any statistical difference in the state of health?

Several large-scale sociological and epidemiological surveys were performed. In one of the most significant surveys, elderly people who had cardiac infarction were investigated for several years. They clearly demonstrated the criterion of their state of health, absolute survivability. All possible factors were taken into account: social status, standard of life, taking medicines, religiousness, etc. It turned out that the level of prosperity and comfort had not influenced the state of health. (The research was carried out in the USA, so there were no starving people among the investigated, but even in the USA the difference in the level of life of various groups of population is quite significant). Religiousness appeared to be an important factor: the believers lived statistically longer. But the most significant factor which surpassed all the others was quite unexpected for the researchers. This was the presence of a home friend, a cat or dog: people who had pets lived longer and were healthier. That's how it's important to have a pet nearby.

In other research, 800 patients were randomly selected in one of the biggest American hospitals, and a group of nuns was praying for their health every day. In a month the progress of treatment of these patients as compared to the control group of people who had practically the same diseases was analyzed. But the control group was not prayed for. The convalescence was statistically shorter in the group which had been prayed for, with all other factors remaining the same.

Such research requires large groups of tested individuals: several hundreds, or even thousands. For some the process goes up, for others - down, but everything is statistically averaged in the group. Classical statistics fully proves itself in such experiments!

In Janet's experiments the calculation was based on detecting subtle individual changes. The methods applied were to follow these changes on all the levels: physical, psychological, and field levels.

When the experiment was being planned I was trying to convince Janet to select people with at least some problems: physical or psychological. Then there would be a way to change for the better. But the persistent delicate lady held her own: she did not want any additional factors to influence the result.

In a month the measurements were repeated, and after processing it was found that no parameters changed. People were healthy and remained the same. This was the first conclusion from the statistics, performed for the whole group of 40 people in comparison with the control 40 persons (within the month 5 people were winnowed out because of different reasons). This was a great frustration for Janet. All her hard labor, sleepless nights at the computer, and long days in the library went to rack and ruin. But the mathematicians decided to stand back and began the analysis on the individual level, the level of each particular person. And such an approach was successful! What an insidious thing is that statistics!

It was discovered that 9 people from the experimental group (prayed for) differed from all the others by a complex of EPC parameters. I.e. those 9 people really changed during that month, while the state of all the others remained practically unchanged!

Skeptic. So where are the proofs that these changes took place as the result of prayer, and not some other factors, say flu or a jackpot.

Author. Of course, the experiment enabled us to disclose only the presence of changes, as compared to the control group. And, in principle, these changes could be caused by various factors, but Janet tried to consider these issues. Her questionnaire had special questions about diseases and important events. 5 people who had such events were eliminated from the study. Therefore, we hope that it was the prayer which was the significant factor causing the changes.

Skeptic. Haven't you tried to measure the influence of prayer directly?

Author. Several times, in different countries. It is especially vivid when the picture is taken before and after the prayer.

Sincere believers demonstrate changes practically always. But the value of such experiments is considerably lower than the value of Janet's study. An individual prays consciously, his psyche and emotions are involved in this process, and he believes in the result. It's impossible to say whether the changes take place because of emotions or prayer. Although both exert positive influence on the state of health.

Skeptic. If everything were like that, the believers would live longer and wouldn't be ill! But there are a lot of them in the hospitals, not less than atheists!

Author. A lot of different factors influence us in our life. Both at physical and spiritual levels. Every minute we take in thousands of bacteria, consume kilograms of nitrates and nitrites, liters of spoiled water, we are exposed to bad thoughts and evil glances. Permanent war of the Good with the Evil goes on. Angels of light always protect their wards from attacks, but the latter themselves trustingly strive after the colored toys of the Kingdom of Evil. And how difficult it is to distinguish a God's messenger from the spirit-tempter! The tempter's speeches are sweet when he lures another victim into his net.

This is why there are so few people living a truly saintly life in an ecologically clean environment. Perhaps, monks on the Solovki or the Valaam Islands in the Far North of Russia or Tibetan monks. But in our life the main condition for health is an optimistic relation to life, correct nutrition, and prayer. If you believe in all this, it works. Well, as you see, now it can be measured.

CAN LOVE BE MEASURED?

The radiance of Buddha shines ceaselessly
Dhammapada

Recently we celebrated the fifty year birthday anniversary of one of our student friends. We drank to him, to his parents, to health, and then he stood up and raised his glass to his wife, with whom he has spent twenty years together and without whom his life would not be so rich and successful as it is. The guests supported the toast with pleasure, drank, and then one of the inveterate single men noticed,

"This is not love, this is a habit".

"It is not true, a habit is to clean the teeth in the morning, when you feel yourself uncomfortable without this, and love is when you can not be without each other for a long time", Andrey, the hero of the event objected.

"There are different stages of love: love to parents, love to homeland, love to children, love to three oranges", Boris entered into conversation.

"Real love is determined by what you are ready to sacrifice for your beloved woman", Andrey commented. He had always had an ability to formulate his thoughts briefly, but precisely.

"But is there a difference between the love for children and love for a sexual partner, although in both cases you are ready to sacrifice everything for their sake, even life," asked Ninochka, taking an elegant sip from a crystal liqueur-glass.

"Love has many sides", Valentin replied, "As the Indians say, 'Attraction of bodies gives birth to passion, Attraction of minds gives birth to respect, Attraction of souls gives birth to friendship. Unification of these three attractions gives birth to Love."

"But love can exist even if these three components are not equal to one another, or when they are different in size, "Shura said. "And love is paradoxical in its essence. We love not for something, but regardless something. That's the nature of loving people, when they don't notice the demerits of the object of their passion they build some ideal image in their soul, which often doesn't comply with reality."

"But the state of love, being in love completely changes the person, this is visible for all people around; he radiates something special into space", Lubasha exclaimed.

"And can love be measured?" Borya asked.

And at that point I recalled the experiments presented at the international conference in Lubljana, Slovenia, by the family of Starchenko 5.

It was a very pleasant trip. Together with Prof. Voeikov from the Moscow University we were invited to this conference by the University of Lubljana. Prof. Igor Kononenko, the organizer of the event, - a mathematician, is one of the European leaders in the field of artificial intelligence. The field of his spiritual interest is consciousness, and its influence upon the surrounding world. As many other serious international scientists, he tries to scientifically substantiate the idea that our life is determined not only by material circumstances, but also by spiritual mood. He had carried out many experiments and finally had settled on our EPC Camera, as a device able to objectively register the influence of emotions and passionate fire upon the human state. Prof. Kononenko had measured the effect of meditation, prayer, Oriental physical and spiritual practices and had statistically proved their objective influence.

Slovenia is a small country situated between Italy, Austria, Hungary, and Croatia. The population is less than 2 million people there, and the whole country can be crossed all over by car in 4 hours. But this small territory has so many beautiful places that it is enough for a month's journey. Crystal mountain lakes with waterfalls and unassailable medieval castles, one of the largest caves in the world with stalactites and stalagmites, gorgeously beautiful underground lakes, and bare blind fish, living in total darkness. A castle standing on the edge above this cave had never been conquered. Once in the Middle Ages the attackers had been standing at its walls for a month and then had met the besieged in the nearby village - they had been cheerfully drinking beer. They had gone down into the cave through the secret passage and had come up through the underground labyrinths on the other side of the mountain. Small villages with durable stone houses, wooden barrels of wine in the cellars, unique for each territory and each vineyard. And the inhabitants consider it an honor if you come to them to drink a glass of wine with house cheese.

An astonishing concentration of talented people in this tiny country. Suffice it to say that the number of our EPC instruments per head there is bigger than in any other country of the world. Several laboratories working in the field of bioenergetics and obtaining interesting results are situated solely in Lubljana. We did not want to miss any presentation at the conference, which happens quite seldom at scientific meetings. But even at the generally high level, the presentation of the Anufrievs from Ekaterinburg attracted everybody's attention.

We had known this couple for a long time. The thin and elegant Elena - a consultant in psychology, living in emotions and intuition, and her husband Victor - a solid businessman, highly spiritual, a reliable and sincere person. They had organized a bioenergetic laboratory in Ekaterinburg, and owing to Victor's business grip and enthusiasm of Elena, this laboratory had become well-known in Ural, and their work had obtained recognition in the world. But let us give the floor to them.

"The aim of the present research is to study the effect of thoughts as the highest manifestation of human psychological activity, on human the psyche. The research was performed with the help of EPC Camera instrument developed by Prof. K. Korotkov, St. Petersburg, Russia. The instrument registers the pictures of EPC fluorescence of fingers which are further processed in computer and a spatial energy field is built around the human body. The capabilities of this instrument enabled us to visually see the changes taking place in the human energy field under the influence of various thoughts.

Research was done on practically healthy people during 2 years. The quantity of tested people: 50. Particularly interesting were the results obtained on close people with quite harmonized energy fields.

Previous research had been carried out by means of a cardiologic diagnostic complex "Varicard" using methods of mathematical analysis of heart rhythm, and it had been found that the thought and heart had been interconnected. The thought, as energy, could have measured the heart rhythm: decrease and increase the frequency of heartbeats, change vegetative balance in the regulation of cardiac activity and influence functional state of regulatory systems of the organism.

A thought of love, coming from one heart to the other in the form of a separate energy cluster, was fixed for the first time in the present research using Gas Discharge Visualization (EPC) technique.

At present many scientists acknowledge that the thought has material nature, that the thought is energy. And this energy is everpenetrating. Numerous telepathic experiments have proven that neither the thickest wall nor long distance has been an obstacle for the thought. From 1919 to 1920's a well-known Russian scientist, doctor-psychologist and academician V.M. Bekhterev had performed a number of experiments on the directed transfer of thought to some distance. First the experiments had been performed on animals, and then on people. He proved scientifically that a person could mentally influence another one. In the 1920's a famous Indian biologist G. Boshe investigated the effect of human thought on vital activity of plants. In the 1930's and 1940's in the West many scientists had been researching the influence of thought and had made experiments on thought transfer (Adrian and Metius, England; Rein and MacDougall, USA; Brunler, Germany). A doctor-psychiatrist from California, Anita Mjul, had demonstrated that thoughts of different quality influence an individual in a different way, changing pulse frequency.

Modern scientific ideas about the world tell about the energy nature of all the living. And the basis of the Energy paradigm is the law of conserving energy. As soon as radioactivity had been discovered, it became obvious that matter is transferred into energy, although gradually. And Einstein's equation E = mc2 gives grounds to speak not only on the transformation of matter into energy, but also on the reverse process - transformation of energy into matter, i.e. proves that energy and matter are a one-in-two manifestation of one and the same universal substance.

The World paradigm of the ancients had always kept the notion about the single moving force of the Universe and the interconnection of all the living in the Cosmos. Different nations in different centuries had called this ever-basic cosmic energy in different ways: Qi in Ancient China, Prana and Vital force in Ancient India. Psychic energy in modern science. "The world process is the manifestation of the single world energy", considered V.M. Behterev, "and wherever the latter was found and in whatever forms, it is manifested everywhere in one and the same correlations and is subject to the same rules and laws". And "this sole world energy is transformed into nervous-psychological energy" in the human being. Thought as the highest manifestation of human psychic activity is also the highest demonstration of psychic energy.

We captured the emanation of psychic energy, coming not from the brains, but from the hearts of individuals while they were sending thoughts of love and good will through space and concentrating on a close, beloved person. (Fig. 42)

Fig. 42 represents the human energy field in initial state (a) and when thought "May the World feel the good" (b) is sent into the space. Activation of heart energy center (heart chakra) takes place. The human energy field picture clearly demonstrates the emanation of energy from the heart, exceeding the width of the general energy field by two or three times, and a mass of emanation from the left little finger is observed on the EPC-grams of fingers. The energy field itself, on the whole, remains even, without breaks, holes, and without emanation from the other energy centers. The energy of love "flies" from the heart. The pulse increases by 10-15 beats a minute, at that.

We managed to capture an energetically isolated cluster coming from the heart of the thought's sender (a) to the heart of the thought's receiver (b) when a significant level of concentration and sincerity was reached during the process of sending thought to the beloved person. On the EPC-grams of fingers, the thought that is sent is located in the heart sector of the left little finger (a, b), while the thought received is located in the heart sector of the right little finger (c, d).

We can distinguish the characteristics of the thought being sent. First, a clear image of the energy cluster, separated from the main energy field, appears. Secondly, the color and form of this cluster is very similar to that of the main field According to the level of brightness, the core of the cluster corresponds to the brightest part of the main aura of the finger or even exceeds it. The cluster may appear at different points of the heart sector and at different distances from the glow of the little finger.

We found the exact moment when the energy cluster of the thought that was sent appeared. The cluster was registered within 1-2 sec, regardless of the distance between the objects (experiments were performed both at 100 inches distance between participants and at 932 miles distance from one city to the other).

Many well-known scientists are preoccupied with the question where the thought is born. The Nobel Price laureate G. Eckles assumed that the brain is the acceptor of thought, and not the producer, i.e. it just accepts and processes thoughts and doesn't produce them itself. Academician N. P. Behtereva considers that this is not absolutely true. The brain does quite well with the simplest thoughts, but when the question is concepts, etc., it is really more complicated. In 1995 a professor of Melbourne University doctor Herst had made a discovery: he declared that the heart had been able to think independently, i.e. to change its rhythm irrespective of the brain. The results obtained with the EPC camera instrument prove this viewpoint. The thought felt deeply by the heart is born in the heart, is transferred by the heart,

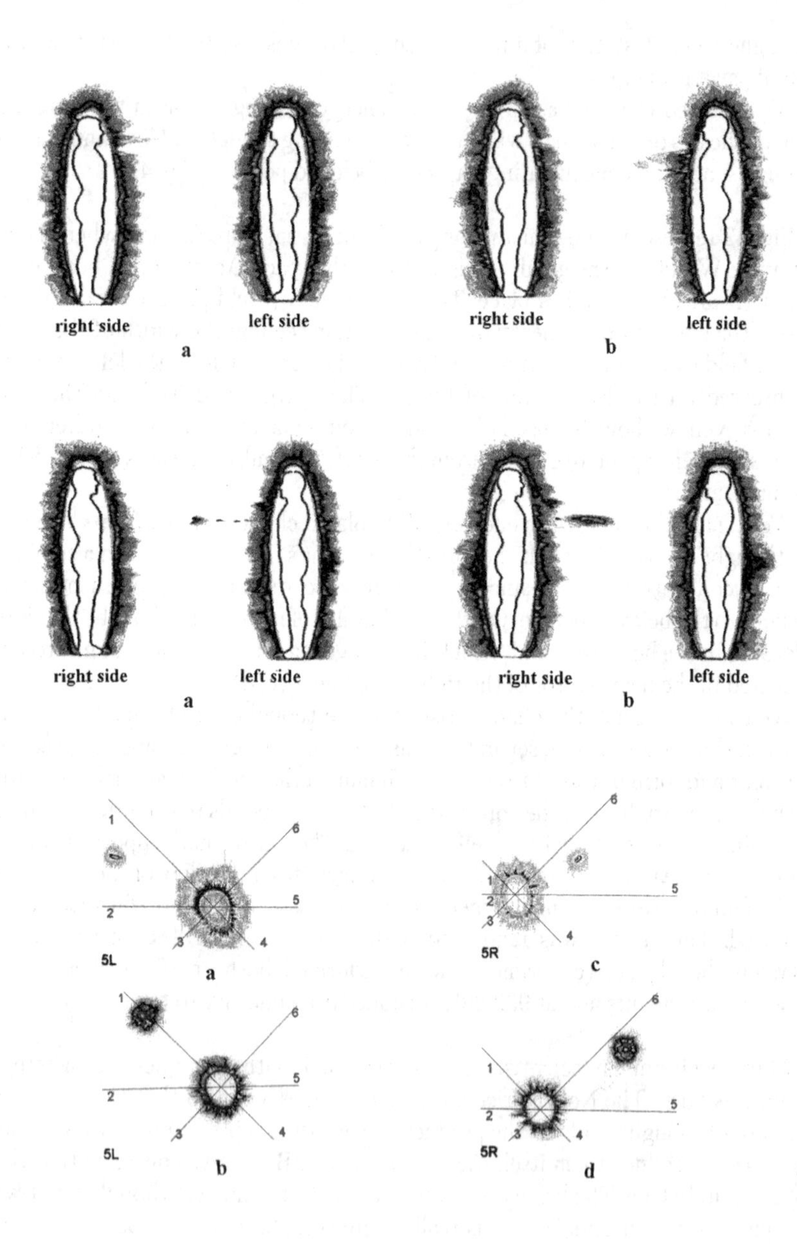

Fig. 42. Exchange of Love between people.

and is received by the heart. And the more heart feelings an individual puts in, i.e. sincerity into the transfer, the more obvious is the increase of pulse. Required heart energy is quasi attached to the thought, so that it has vital activity. In the Studies of Live Ethics, which came through Nicholas and Helena Roerichs it is said, "It is not the brain substance that thinks. Time has come for all to recognize that thought is born in the fire centers. Thought exists as something weighty and invisible, but it is necessary to understand that matter is not the only thing that matters. The thought is sent through the heart and is received through the heart, as well". Thus, for the first time, we registered the thought of love, coming from one heart to another one. Human thought is energy, which can influence plants, animals and the human being, that is, all the environment. In accordance with the law of energy conservation, the thought is indestructible. Mankind is constantly thinking in the process of its activity, making its contribution to the energetics of earth. And this means that each person should realize high responsibility for the quality of his or her thoughts.

Isn't it a wonderful research! Moreover, it's worth mentioning that the phenomenon of sending "love energy" has been registered many times, with different people, sincerely loving each other and sending their tender feelings to one another. So, it turns out that Love is really the only force that can withstand Evil and save Humanity from the coming catastrophes.

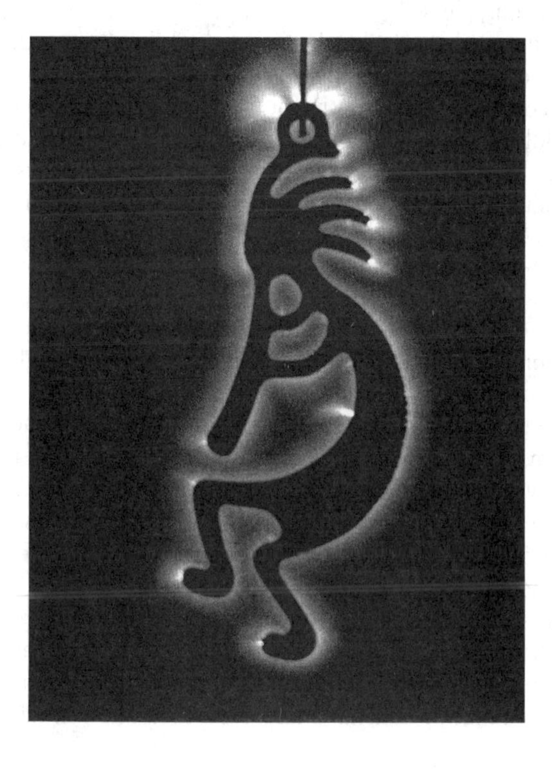

QUEST FOR ADVENTURE

"OK, guys, the helicopter can not fly further. If you wish I will leave you here", said the commander of the helicopter and pointed at the white foggy wall covering the ravine. It was obvious that even our Russian helicopter crew would not risk in such conditions. What was more, at 4800 m (about 16000 feet) above the sea level, at the very foot of the Chomolungma, the highest mountain in the world.

"And is it far to the cabins from here?" asked one of us.

"Not really, maybe a couple of hours", the helicopter pilot replied cheerfully.

As it was found out later, he had been estimating the distance from the sky, and what had seemed to him a straight road curving along the mountain river, as a matter of fact, came out to be a long monotonous track slowly moving from one gully to another one. Pilots and pedestrians have slightly different notions about the length of the way.

"Well, then we're going out".

The helicopter nodded down, made a circle turn, and in a few minutes hovered in a meter from the glade covered with snow. The pilot hospitably opened the door of the cockpit and invited:

"Who's going out, please!"

We rapidly dropped the rucksacks and jumped down, onto the snow whirling from the lift rotors, one after another. A final wave at parting, and the buzz of the rotary-wing machine moved to another heaven plane. The three of us were left in the fog, in the middle of the Nepal Mountains, having absolutely no idea where the top was and where the bottom was. I sadly glanced at the pile of rucksacks. The powerful helicopter should have brought them to the base camp. Their total weight was approximately equal to ours. It was evident that was beyond our strength.

"We take the equipment and most necessary things," I announced to the freezing team of the two young pretty girls. "We'll hide the rest".

"For keeps?" one of the young girls asked gingerly.

"No, we'll come back later and get it", I replied, and with a gloomy thought to myself, 'If we come back.'

The most important thing in hiding the bags is to make it in such a way that you could find them earlier than anybody else. When you go bottom-up, the landscape looks absolutely different than when you go top-down. Therefore, the process of searching for the place, with risky jumps on the wet slippery stones, carrying over the bags and camouflaging them, took a couple of hours. The girls already got absolutely frozen. We shouldered the rucksacks and dragged up.

In two hours it became clear that the pilot had strongly underestimated the way. We got onto the track, but it was still a long way to the base camp. Up and up. I cursed myself under my breath that we had not hidden the equipment; although I clearly understood that I would have never done that. The altitude approximated to

5000 m, and it was obviously not enough air to breathe, and we had too little time for normal acclimatization. What a steep mountainside, each step came difficult, and the damned rucksack pulled the shoulders back.

The mountainside went sharply upward. Probably, we should turn somewhere. I looked around and saw pink ribbons on the stones to the right and to the left. So, that was the way. Slowly moving the sticks, we reached those stones, passed them, and a small hollow opened up before our eyes. When I saw it, I put down the rucksack and fell on it with relief, looking around.

We came out onto some sacred Tibetan place. This was a flat hollow, a few hundred meters in diameter. From one side the slope went sharply up, from the other - fall to the glacier. Stone plates 1-1.5 meter high with letters on both sides were set on the horizontal ground in concentric circles. There were a lot of these plates, maybe several hundreds. Inscriptions were carved on huge natural stone lumps standing in the hollow. Between the huge lumps, from one side of the hollow to the other, the ropes were stretched and sheets with sacred texts were hanging on them. Many lost their color and contents with time, the other were absolutely fresh. The picture produced a mystic impression, resembling decorations from a fantasy film, but not a piece of real landscape.

I imagined how much effort and energy would have been spent in order to first carve all these inscriptions, and then drag these stone lumps here, practically to 5000 m above sea level (more than 16000 feet). There were no settlements higher. Only camps of alpinists and tourist pilgrims. What was that - a place of sacrifice to numerous gods, evidence of redemption of sins, or a pagan temple for holding secret rituals? And, maybe, as it is popular to prove in modern scientific treatises, that was an ancient observatory, enabling to determine a direction in the constellation of Lira and the Pleiades at the night of the equinox? In order to start from here and go somewhere far off.

From the landscape my thoughts passed to the EPC images of the local people. Why were they so different from ours, European? And what were the EPC-grams of monks from local monasteries like? But EPC images of a Tibetan living in the USA did not principally differ from those of Americans and Europeans! So, what was that: the effect of the altitude, hard labor, or nutrition? Next year we will need to plan a special program of investigating the local people, particularly, monks. At different altitudes, from Katmandu, the capital of Nepal, to transcendental settlements.

By the way, I should not forget to mention the Indians. Very interesting data. And about the Olympic combined teams. Would be good to take EPC-grams of Olympic Champion Alexander Karelin again, as he is coming to defend his doctoral thesis. A conference on live and dead water in Germany in December. Should necessarily participate, there will be an interesting company, all world leading specialists in water research. Yes, many different directions, and one is more interesting than the other. But these are topics for another book. If time permits, will finish it in a year. With God's help and if everybody is healthy. And wish you the same, dear readers.

I looked at the resting girls. Seemed they slightly recovered. We could move forward. We stood up, reluctantly got under the rucksacks resting on the stones, becoming heavier and heavier with every hour of our way, hardly drew ourselves up, and dragged further. The fog suddenly dispersed, and the magnificent Himalayan 8,000-meter - 28,000-feet Mountains appeared in their prodigious beauty right before us. The track came sharply up.

Fig.43. Good bye, helicopter......

Fig.44. On the way to the Summit.

Fig.45. Traces of ancient wisdom.

Part II

SPIRAL OF LIFE

PROLOGUE

We are tearing at full speed in a rented "Chrysler" in the vast expanses of Arizona. Me and my friend. Somewhere between Los Angeles and Phoenix. The road, straight as an arrow, is flowing right into the skyline. Views of blossoming spring desert are changing around. Light jazz plays. Calm, melancholy mood. We continue our leisurely road discussion about different stuff, and suddenly the friend asks,

"Konstantin, have you ever thought why you were the one who managed to achieve something in the research of biological energy fields? Thousands of people studied the Kirlian effect, invented several variations, but only you could turn this semi-mystical method into a serious scientific technique, applied in medicine, sport, and industry. You are traveling around the world, presenting at most prestigious universities, although yet ten years ago all this pertained to the category of semi-scientific curiosities. Why?"

I start pondering over it - why, really? And, thoughtfully stroking my bald spot, I tried to discourse on this topic, to philosophize; it was good that the friend drove with confidence and there was enough time on the way:

"Of course, there are a lot of reasons here. The main reason is that the time was appropriate and public consciousness developed well enough to understand. A new class of equipment appeared which was inconceivable yet five years ago. New methods of processing information. As to the achievements of our team and much interest to our work - this is mainly governed by the style of our approach both to work and to life. The fact is that in all our activity we follow the main principles of alpinism, manifestly or implicitly. We still have a long way to go, so I will try to formulate these principles.

When you approach a mountain, you see a giant in front of you, going up into the clouds, with steep walls and threatening glaciers. What is a human being compared to it - an ant, a small insect, a speck of dust on an elephant's leg? If once the mountain moved, or a hurricane sprang up, - the speck of dust would be swept away, carried away by the gust. It becomes terrifying, and the first impulse is a desire to go back home. But it is indecent to give up, so you make the first step, then the next one, another one, - and so, little by little, you overcome the whole way and find yourself on the top. Therefore, the first principle:

I. Forward and up step by step.
Do not be afraid of the task facing you and make the first step.

And you are on the top. The goal is reached. Nothing is higher than that. You feel joy and satisfaction that you were able to do it. But this is not the end. The top is just a milestone on the way. Maybe not the most important on your way. Reaching the top is not important, the way it is done is.

But why? Isn't reaching the summits the main purpose of alpinism?

This is both true and false. Several difficult paths can lead to one and the same top. Everyone chooses the way according to one's capabilities and proceeds, often risking his or her life. And, especially in the Alps, it frequently happens that after ascending a steep wall you get to the summit and see a mountain cabin where people come via a funicular railway to drink coffee. You come up, staggering from tiredness, all in dirt and with grazes, vestiges of the lightning flying by around you, and see civilized burghers who climbed up the same summit by a cable car, rocking their fatty bellies. And, as soon as you step from the rock to the tile floor, the difference between you and them disappears. Here follows the second principle:

II. The goal is nothing, only the process is important.

But so many people set aims in their life, strive for them, spend life in order to reach these aims. What about the principle "The end justifies the means"?

You are absolutely right noticing that they "spend life in order to reach these aims", often destroying everything and everybody on their way. But here the aim is reached. And what? A person understands that the life has lost its meaning. Then - depression and frustration. There are so many cases when the inveterate gamblers, having cleaned up a big profit, blow their brains out - life lost its meaning for them. Many professional sportsmen fall into depression by the end of their sport carrier - they can't find their place in life. In alpinism, standing on the top, you see the endless mountain country, you are attracted by new horizons; you already imagine another summit; and you remember that the way down lies before you, and it can be even more dangerous than the rise. Here comes the third principle:

III. Everything passed is in the past, only the future is ahead.

Having reached the top it's important not to fall into euphoria after what was achieved, but remember about the next step, next summit, and see perspectives of development before you. This idea is shown best in the notion of traverse - a rise when mountaineers move along mountain ridges, clambering from one top to the other, evading descents to the valley. Some summits might be higher, others - lower, a descent is inevitable after a rise, and it is like a metaphor of our life. It's impossible to always go up.

The forth principle:

IV. A rise is always followed by a descent, a victory by a defeat, remember this and do not lose courage.

But aren't there dead-end situations, from which one can't get out?

People often consider a simple change of conditions, which frequently comes with difficulty, a dead-end. Each situation ends up with something, and in the next process, a further development starts. Most difficult situations might be trials sent to us from the above. There are many examples of that in the Bible. Many people are not used to overcoming or suffering troubles. They panic at seeing the first clouds in the bright sky. But the worst outcome of the most extreme case is Death. This is the highest point, Culmination, always unexpected and always the inevitable finale of the melodrama of life. But this is not the end, not black thick darkness, this is a transition through a threshold, exit to another level of existence of the information body, the human soul. So, there are no dead-ends. There are complex situations, blind alleys, which seem to be the center of all dark forces of the Night, when it appears that all opportunities have been attempted and no more forces have remained. In such moments in life it is often necessary to just hide yourself and wait. Wait for the first rays of the sun. Rely on the God's Will and give yourself to the succession of events. At the same time, the following paradoxical principle is known among mountaineers of the top class:

V. The worse, the better.

What could this mean?

When you ascend on a clear sunny day, easily overcoming rocky bastions, and come down to the camp by dinner time, cheerfully knocking the boot with an ice-axe, after some time has passed, nothing remains in the memory from this day. If a thunderstorm broke out on the way, with lightning strokes to the ridge and the ice-axe buzzed with the electricity, if you got up to the summit at your last gasp, gulping rarefied air at each step, if you had to spend a cold night sitting at a small mountain ledge and hoping only for the ropes bound to the irons driven into the rock - that remains in your memory forever, for the rest of life. And all our life is just the today plus memories.

What a principle! In this way, one has to create dramatic situations, and then be happy having experienced those!

Well, not so bad. The life of every person is full of problems and troubles anyway. One shouldn't create them. If you have no problems, this means you are either sleeping or already died. Troubles come themselves. "It never rains but it pours". It's necessary to create the moments of happiness and guard every rare drop of happiness given to us. They are so infrequent in our life! Only idiots are always happy.

But what shall be done if a person tries to make, to create something, and nothing ever comes out. He takes up one thing, another thing, and is always a failure. Why are some people so lucky to achieve everything and others have just a pell-mell of troubles?

This is, of course, a question for long discussions. We will come back to it. A couple of words won't be enough. First of all, much depends on whether a person does everything for other people or for himself. The more you give, the more you receive. The stronger you espouse to the labor of love, not thinking about recompense, the better the results. Most unlucky fellows are digging their small pit, never thinking if somebody else needs it. What is more, we are all different and everybody has one's own level of capabilities in one's sphere. We are gradually developing scientific methods of revealing optimal fields of activity for a given individual. The number of such approaches increases. And we can simply say as follows: you can start any activity and spend three years doing it. Writing books, pictures, traveling, science, or business. Then you need to stop, look around, see what comes out and understand if it is yours or not. If not - it might be better to step back and look for another way. Each alpinist has unconquered peaks, and at each huge mountain there lie mountaineers who did not give up. Here follows the sixth principle:

VI. Stop, look around, - and make a decision.

Thus, a whole life will be the searching and looking around!

Of course, we can recall a lot of people who served all their life to one idea or one Muse. But we can name not less examples of great people who forced themselves in different spheres and made their significant contribution in some or several. Patience and persistence are important. Look how a beast can be on the watch of prey, hiding under cover and almost not breathing. On the wall rise, standing in an uncomfortable pose, you have to wait hours until your mate passes a difficult part and fastens the rope. Once, for three days in the mountains, we were waiting until the blizzard was over, lying in a small tent, buried in a snow cave on the mountain slope, without food, without an opportunity to move up or down. We were just lying, playing cards painted on toilet paper, and listening to the wail of the wind, subdued by the walls of the snow cave. Our talks revolved around memories of fried potatoes with bacon and beer. And every hour a loser crawled to dig out the entrance, in order not to be smothered under the mass of snow.

If you are able to endure and remember about your goal, if you calmly wait for your hour - it will certainly come: the hurricane will end, the black line will pass, and the sun will look out again. You need to be patient only, and persistently, notwithstanding any obstacles, move to your dream. As a rhinoceros - paying no attention to the obstacles and not giving up when misfortunes come. And this is one of the main life principles:

VII. Patience, patience and persistence in moving.

"Sorry", exclaimed my friend, "From all these principles I haven't gotten the main law of your life philosophy. In Buddhism it is "Life as a continuous wheel of suffering," and one has to find the way out from the chain of reincarnations. In paganism this is attaching gods to oneself. In Christianity - serving God. And you? What's the basis of your relation to life?"

"This is the main principle of my philosophy. It may be formulated as follows:

VIII. Life is a Wonderful Journey.

It gives us for happiness and grief, where we, are actors, authors and producers! We need to be happy for the good which we are given, and bear the rough and misfortunes well.

Then we are doomed to Success!

This discussion could have continued for a long time, but a wayside sign-board of a "Best Western" motel came into view in front of us - this was the place of the next stop of our many-days journey. I checked the map again, and we turned to "Exit 75" from the highway, leading to a group of buildings with motel, gas station, and a restaurant with synthetic American food. Dinner, shower, and getting to bed; and then another 250 miles on the following day to the place of the lecture.

And so, after this journey, as after many similar talks with different people in different countries, I came to this idea to write a book, - an essay about life, meetings, work. And illustrate it with situations from the alpinist practice. There were a lot of them.

This book is not memoirs, not the expounding of philosophical ideas; this is rather a reflection of the powerful process of collective development, illustrated by a particular fate. Don't look for concrete facts, dates, and names. This is a collection of essays. Multicolored stones strung on one thread. A thread of an individual fate in a critical epoch of development for Humanity on the border of two Centuries, two Millennia.

I. Forward and up step by step.

II. The goal is nothing, only the process is important.

III. Everything passed is in the past, only a future is ahead.

IV. A rise is always followed by a descent, a victory by a defeat - by defeat, remember this and do not lose Courage.

V. The worse, the better.

VI. Stop, look around, - and make a decision.

VII. Patience, patience and persistence in moving.

VIII. Life is a Wonderful Journey.

BREAK POINT

It is a bad sign when things go well for a long time - expect trouble!

In summer 1983 a group of mountaineers was ascending to an unconquered wall of the Central Pamirs in Tajikistan. The wall was situated in the spurs of the Gissar ridge in a distant canyon. One could reach it only on foot, tracking a steep path. Therefore, it hadn't yet been passed through. We were filing past for hours, making our way through steep tracks in dark ravines, we could hear the noise of mountain rivers somewhere far beneath under our feet, and the burning hot sky appeared above. It was burning so hot that it seemed everything around was melting. You go and go, and it looks like there is no end and no beginning. And suddenly narrow walls of the canyon open and a green valley with streams and small waterfalls appears in front of you. Mountain ridges stretch on the sides of the valley, and an entirely steep wall rests in the end, where the valley expands slightly moving upwards, as if blocking it.

Such walls are not uncommon in the Pamirs. Many valleys stretching along heavy rivers close with steep walls unexpectedly, when you rise higher and higher into the mountains through these valleys. It is connected with the peculiar structure of the Pamir Mountains. Young growing ridges, they rose protruding from under the ground because of the powerful tectonic forces. In some places the plates stood on the edge, creating the basis for the mountain ridges. In others - the earth's crust was breaking in small fissures, and the mountain chains dispersed.

Such walls are a Mecca for climbers. They represent a whole set of mountain difficulties. They are usually steep, hundreds of meters tall, and sometimes several kilometers long. And, as a rule, they end up with an icy cupola, leading to the very top. The ridges contain good, solid rocks, like granite, quartz, or marble. Climbing such walls affords a lot of pleasure. Therefore, most of the walls located near the tracks are known and often visited, and a large number of routes are set there. It is a great challenge to find a new wall! And matter of luck. Such places are practically indistinguishable from a helicopter. Moreover, no pilot will be looking down and paying attention to the "small" peculiarities of the mountain relief. So, in fact, the new objects are usually found by the mountain tourists moving along the Pamirs' canyons, or by the frontier guards walking the rounds of the guarded territory.

The wall we approached was one of such walls, absolutely unknown to the sport community. The wall, promised if not golden medals but at least some prizes in the Championship of Alpinism in the Soviet Union.

Going out to the valley we stopped, threw off the weighty rucksacks and began to examine the wall through binoculars. Then the most difficult part of work started: it was necessary to carefully investigate the wall and plan the route to ascend the entire wall.

We were eight. An assault detachment of four people would go up the wall, the other group staying at the bottom of the wall in a tent camp set up on the moraine - a pile of stone lumps left in a chaotic mess by the receding ice stream. The four at the bottom were to play cards for days and each five hours establish radio contact with the assault detachment and with the base camp.

In four hours we approached the very bottom of the wall and chose a perfect place for a bivouac. That was a relatively flat place amidst stones. The whole group started setting tents with pleasure. We spent the next day investigating the wall and discussing possible ways to ascend.

In the morning, after a hearty meal the main group moved up. The first parts of the way turned out to be easy - these were scabbings, small ridges, and little walls. It was easy to organize rope belays, so we moved rather quickly. The first one climbed with a light backpack and a heavy armor of climbing gear, ropes, pitons, carbines, safety hooks and hammers. He was looking like a Medieval knight, fully covered with weapons clinking at each step. The second pair pulled the main luggage, climbed with ropes, dragged the rucksacks with tents, food, and stocks of equipment. Ice slopes were waiting for us in the upper part of the wall, so we had to carry ice pitons, ice-axes, and ice hammers with us.

Each 10-15 meters the leader drove in an iron or fixed a special metal piton in a crack in the rock, put a rope through a carbine and snapped it up. This provided belay (security) for him. In case of a sudden fall he would have just been hanging by the rope. But on the whole, all guys in our group were well trained, full of strength and energy, and were not going to fall down.

Six ropes were passed in the first day, i.e. 240 meters (800 feet) of the vertical rock. Mountaineers consider it a reasonable achievement. By the end of the day the leader found a little ledge, where a small tent could be set. A tiny brook was purling in the rock not far from us. This was all needed for life. A place to lie down and water to make food.

Investigating the wall on the previous day, we saw that the stones falling from the above had passed by the lower part of the wall, not even touching it. This indicated that there was a part with a negative slope in the middle of the wall, i.e. the wall was as if hanging above the moraine. All the falling rocks were passing by far from the lower part (and they inevitably fell as the rocks cracked from temperature differences, when the glacier and snow thawed on the top). Indeed, in about three hours we heard the stones hurtling down aside. Nobody paid attention to this. We all gathered in our bivouac, happy and satisfied with a successful day. We had a soup and mixed puree with dry meat, and ate with pleasure. We drank tea, and everybody settled to sleep as we could.

The first half of the next day passed in the same successful way. The climbing was difficult, but very beautiful. By 2 p.m. the group had passed another four ropes. After that we got to a place where the rock impended over us vehemently. Where before we could have seen the sky moving over our heads now only beetling-over rocky balconies were seen above. Here two thirds of the wall, about 300 meters (1000 feet), was occupied by the impending part. And then, higher, there was a flatter part.

Rising across these impending cliffs, we almost cut off our way back. It was impossible to come down from the overhang, as the end of the rope put down from any point of this beetling over "belly" was swinging in the air several meters from the rock surface, and it was practically impossible to catch it. Only in Hollywood moves can Tom Cruz fly over from one rock to the other in a jump. This is impossible in life. And the distance to the moraine below was already too long - about four hundred meters. But we were far from psychological problems - the energy boiled over in the young organisms and nobody had any doubts that we should move upwards.

The weather was perfect, not a cloud in the sky. We were at the peak of sport training. And the main thing - we were united by the certitude self-reliance that together we would overcome any technical difficulties. We had a great team by that time!

However, our speed abruptly slowed down. There is a big difference when you are climbing a steep rock and you body is moving vertically, and when you are climbing a beetling over a balcony hanging almost head first. This is the highest class of the alpinist technique, which requires dexterity, a colossal sense of balance, great powers of endurance, and special equipment.

Here Alexander came forward. He was the most puny and thin in the group, but everyone knew that as to passability, i.e. the ability to pass technically difficult places, he had no match. He could catch on to any irregularities and hang head first with no time limit. Having a relatively small weight, he had powerful hands and feet. He put on all the equipment and started climbing up. Everybody understood that just a few mountaineers in Russia could do this with such virtuosity and grace. Perhaps, it was worth fixing safety points more often, but he was self-confident: every step was absolutely proven and safe. Moreover, if a part of the balcony would have suddenly fallen off together with him, which sometimes happens during such climbs, he would have swung down and hung on the rope which still did not reach the underlying slopes. Therefore, he was climbing calmly, without haste, but at the same time constantly moving upwards. In about 5 hours of the most difficult climbing he reached the top of the rocky balcony and found a wonderful place for an overnight stop. A broken off cliff formed a ground full of stones. After cleaning it a little bit we could put a tent there. That place was so convenient that it could be compared with an overnight stop at a usual moraine. Alexander took up his mate, they fixed the rope, and started preparing a night's lodging.

The second pair rose in about two hours, seriously fagged out. All the psychological risk of passing difficult parts fell to the first one in the pair. The second one had to drag weighty rucksacks with luggage and reserve equipment, drive out all the irons and gather ropes along the way. Therefore, all participants of the group were important. Exchanging jokes and discussing small details of the rise, we settled a bivouac and prepared for sleep. Of course, that sleep was rather far from the common human rest. Each of us was wearing a helmet and in a full safety outfit. Each was fastened to the rope stretched along the rock. So, if a small earthquake had occurred, this should not have led to tragic consequences. Such conditions were quite usual for mountaineers.

And the general picture was grandiose! In the eyes of an eagle soaring above the canyon people on a steep six hundred meter (2000 feet) long wall seemed to be small swarming insects. They were like ants clambering the wall of a five-storcy building. The difference was only in the fact that an ant can not fall down from the wall, and mountaineers are supported on it only owing to their hands and dexterity.

With the first rays of sun we began getting ready to start. On such a small ground it always takes quite a lot of time. In an hour Alexander started climbing up while all the others continued packing the camp.

The circumstances developed favorably. Notwithstanding the fact that the sun was lighting the wall practically all day long, it was rather high, so the air was cool, and this gave an opportunity to climb, feeling special pleasures.

The group was climbing up through the whole day, not very rapidly, but confidently moving to the top. In about four hours Alex passed the next beetling over balcony and saw the sky beyond the overhanging ridges. This meant that the steepness of the wall started to decrease above. "Done, we have won", Alex thought, and shouted at the top of his voice, "Guys, I see the exit!" It became clear that the most difficult part of the wall was passed. Perhaps, there was only one rope left to reach the top. And then the usual complex, but customary technical work started. However, the last part of the wall happened to be extremely difficult. The rocks were smooth, with practically no hooks which we could catch with hands and to stand up. The rock was still overhanging, and several times Alex thought of starting to drill the holes in the rock and hammering in special irons in order to hang the rope ladders. But then he decided to give up the idea, as after several attempts he still managed to find some hooks in the rock. However, the progress was very slow. By 6 p.m. he did not climb a full rope distance from the place of previous support.

It would be getting dark in an hour. An absolutely smooth wall was overhanging above him. Thus, having noticed the first crack Alex drove an iron in it, hung a rope, and came down carefully to the place of previous support, where his mate was waiting for him. "We will stay overnight here." - He said. - "The place is quite safe; and in the morning we will start moving further. Very little is left." The place was really comfortable for staying overnight. We were standing on a rocky slope, rather large and wide, from the alpinist viewpoint, of course. This meant that four people could stand there, or, which was the same, a tent could be set up and we could spend overnight sitting, suspended on ropes. The rock was still overhanging above, so we could feel ourselves safe from the falling stones and lumps of ice, which were hurtling down for many meters from time to time.

The place was located as if in a small ravine. Steep battlements rose along the edges - mountain cliffs, protecting the place from sudden gusts. In short, the place was convenient for an overnight stop. Pieces of ice were partly coming into view up on the nearby battlement. There was no running water in the evening at such a height, but within half an hour on a primus stove, the ice turned into fresh water.

The guys pulled up the second pair and settled a bivouac in an hour. Drank tea, ate simple broth. And then, having wrapped ourselves up in downy coats and having exchanged jokes, we were getting ready to sleep. We were in good spirits. It was clear that tomorrow we would be on the top. This did not cause any doubts with Alex's skills. However, a few hours of active work were required. But this is quite routine work, although for any other alpinist this wall could have become absolutely impassable. We already discussed that tomorrow we would pass the overhang in half a day and come out to more flat slopes and at any rate by night would be under the cupola of the summit. And this meant that there were all chances to come down to the base camp in a couple of days.

It was total darkness. Having discussed final plans, we turned off the small lamp and went to sleep. The next day we would have to get up early. Stones and ice lumps were hurtling down on the other side of the tent, but all that was in another world.

We almost got to sleep when we heard a sudden crack, a bump and somebody's groan. "What is it?" someone cried. Turning on the lamp, we saw Alex sitting in the sleeping bag, clasping hands to his bosom and shaking in pain. A large lump of ice was lying on his lap, and a big hole gaped in the tent section. As we saw later, a huge ice lump had fallen from the above, split into many pieces, and one of them had ricocheted off the rock and fell into the tent. We cleared a space on the floor moving to the walls, laid Alex down, and doctor Igor started to examine him. When he touched his chest, Alex cried.

"Seems that a couple of ribs are broken," said Igor, "And the clavicle," he added after the next grimace of pain. He moved his ear to the patient's chest and started listening carefully.

"The breathing is clear. Thank God. Will hope he's got no internal injuries. Looks like a serious injury with fractures. Three days of bed rest with chicken soup and a pretty nurse."

The last words were a usual joke in our company. It was especially nice at the height of 4700 meters (15,500 feet), and at a small rocky bench under the darkly overhanging bastion. The leftovers were only tea and some chocolate. According to our plans, we should have already been on the top.

But many a true word is spoken in jest, and Alex really spent the next three days "in bed," lying on the same bench, and we were persistently making our way up to the ridge. An iron, a ladder, another iron - what a hard and exhausting work it was! The rock was so steep that a couple of times the leader fell, hung by the rope, and the next one came forward instead of him - one by one. What was more; it started sleeting on the second day, which "enriched" our 'pleasant' feelings.

Thus, all our forecasts and plans were dispelled in a split second: we thought that we need just a couple of hours to get to the summit, but Divine intent placed everything in its own way.

There is such a notion in mathematics: a break point, when a smooth curve suddenly makes a jump and continues at another level, which seems to have no connection with the previous one. This often happens in life.

A rescue party met us on the ridge - with warm clothes, food, and hot tea. All working together, we pulled Alex out.

From then on, we started using a new rule: never be happy in advance. Nobody knows what will turn out.

PRACTICAL PRELUDE

It was getting close to midnight. The three were sitting around the table and waiting. The tea was already cold, dried spice-cakes were lonely lying on the plate. This was the Board of Directors of the Company. As you understand, any self-respecting company should have a Board of Directors.

The door opened, and the Boss entered. Everybody started moving, turned on the teakettle, and somebody brought the Boss' big mug.

The First Director began the meeting.

"Congratulations, my friends. We have been creating the new line of research for five years together: EPC bioelectrography. Five years we have been living on dry rations and have been working 14 hours a day with no weekends. We have denied ourselves everything and have spent all profit for the development of scientific work, preparation of new programs, materials and books. We have passed through our children's diseases, internal crises, hopes and frustrations. In the course of all this time we haven't had a penny of investment, and all projects with partners and investors have finished in disappointment. But we have continued to persistently move ahead."

And so, finally, when we have created all this as a system, when it became clear what, how, and why everything works, we have been tricked for the umpteenth time.

"Who, our former partners again?"

"Of course. We create partners, and then they steal something from us and become our competitors."

"Well, who this time? Really, Pushgorod?" the Second Director asked.

"Certainly", the First Director replied. "The instruments we developed together which they built according to our old scheme, and for which we spent a lot of money and effort, are now sold by our former partners."

"But their instrument is significantly worse than ours. We still can not make the samples received from them work properly," the Second Director exclaimed.

"But they are twice as cheap as ours," the Third Director commented darkly.

"But where do they get the software? This piece of iron does not cost anything without software. We have been creating this software for many years already and it can't be quickly reproduced. If 9 pregnant women are taken together, they can't give birth to a baby", the Second Director exclaimed.

"The software is just being stolen," the First Director replied, "It's typical for the Soviet way of living". As you realize, it's no trouble for a talented programmer to break all our security protections. Pushgorod took a very convenient position: they sell their product to anyone without software, and what happens next is not their business."

"But it is unfair!" the Second Director cried in a temper.

Everyone burst out laughing. But here the Boss who had been listening attentively to the discussion broke into conversation,

"My friends! As for me, this is an occasion for a holiday. This is the best acknowledgement of what we have created! It means that people realize that the technique is working and have started searching for ways to use it in their practice. We together have created a new scientific and technical line of investigation: investigation of energy-informational fields of biological subjects. There was nothing of this kind 5 years ago. Exotic Kirlian photography existed, and serious scientists thought very skeptically about it. And they were right. Because in the modern world it is not enough to announce something, it is necessary also to prove it. We have managed to do this in these years. We have been led by the understanding that we have been creating a new method for Humankind, a technique for seeing the invisible, to observe the lacy dance of the human energy fields with one's own eyes. Look how many articles on EPC bioelectrography have already been published. Next year a text book for students of "Biomedicine electronics" will come out. Hundreds of people from all over the world visit our annual Congresses. How many countries were present this year? 23! And congresses and workshops in different countries! Each year 8 to 10 trips minimum. And we are invited, everything is paid, big audiences and mass media gather. Already this year: Spain, France, Italy, Germany, and tours around the USA; Korea in autumn, again the USA, Canada, and Switzerland are planned. Our trip to Barcelona has become the beginning of official relationships between St. Petersburg and Barcelona.

"And all this because we demonstrate how the informational field lives, how it changes under diseases, in the process of treatment, in different life circumstances. And another thing, very important. We develop the approach to scientific investigation of spiritual processes. "Science. Information. Spirit" - this is the general name of our Congresses and the general slogan of all our work. Material tendencies are being actively developed in the world: genetically modified products, cloning of biological creatures, right up to the human being. As a backward wave, as a counterbalance, craving for eternal values and spirituality increases. Many people in our country and in the whole world strive for the spiritual, but everyone is already bored with the fruitless discussions and frequent repetitions of one and the same. Many people in the world are interested in real results."

"Remember, we always say: life is the balance between the material body and the eternal Soul. And our research, our developments, our devices can translate many hidden processes into modern language and visualize them."

"We have been aiming at making our instrument, software, and techniques maximally easy and available from the very beginning. This comes out well, although not so quickly as we would wish. Too few resources. But we move forward and prove perspectives and practical value of the developed ideas. The technique becomes more and more widespread and acknowledged. This resulted in competitors and imitators. Nothing surprising and nothing frightening."

"But we can be fully replaced," the Third Director said quietly.

"Yes, we can. If we stop. Look how the biggest computer companies work:

Microsoft, IBM, Hewlett Packard, etc. They constantly offer something new. Develop all the time. And the offers are so interesting and attractive that the user can't reject them and eagerly offers his money like a lamb."

"But we are not Microsoft," the Second Director exclaimed, "And what output we have on the whole... Each quarter we search how to cover financial holes."

"And the circle of our consumers is still considerably smaller. But this is for the present time. A few more years of development, and our techniques will be widely introduced in medicine, sport, and occupational selection. And, again, we should prepare certain steps for this. How is the compact instrument?"

"At full speed," the First Director replied. "The first instruments are planned for November."

"Excellent! This will give another impetus to the wide distribution. And I had thought and decided that we will not be securing our basic programs. Let people use them. If they are honest they will pay. If their level of consciousness is not high enough - God will punish them!"

"For how long can we endure this? ... " the Second Director muttered.

"This is not our question. Beyond our scope," the Boss said. "Remember, I have always told you: our main goal is to do the Work and move forward. With clear honest eyes. If someone chooses crooked ways - it's their problem. There are global laws of development, the divine commandments given in the Bible, and mainly for us - peace in the Soul and a good conscience. And God will serve justice for everyone!"

The Boss toyed with his mug of cold tea and meditated on something for a minute. Then he continued,

"As to Pushgorod - they are our colleagues, in the past they helped us immensely and we have always proposed joint developments with them. Unfortunately, they can't provide the necessary level of quality and service. There are too many problems with their instruments. Therefore, we don't organize joint work, and don't reject them. They are wonderful people and do everything as they can and on their level. It is always better to keep friends than rise against them. Now, however, we have a stimulus to further active development."

"So, we start a new stage of our activity. I officially announce: 19 working hours a day from now on, with 10 of that in our dreams. Only this way can new ideas be born. And the main thing, enjoy all activities and maintain one's good health. What about a trip for hunting mushrooms on Saturday? It is said the mushrooms have grown after these rains. After the forest, we will have time to sit and discuss our plans. It's too late now. 1 a. m. "

"Soon the bridges will be drawn," the Second Director put in. "So, we will be uselessly sitting up all night long. "Well, let's go home. And I ask you to consider all we have discussed while sleeping," the Boss decisively stood up.

Everyone got up and started moving. The haze of the Petersburg White Night was flowing behind the window.

POWER

Most part of my life passed during the Soviet power. When we look at this period now, it seems to be a little bit improbable. What was that communist regime? What was it for? Why was it in Russia that it captured the masses and was able to keep all the country in poverty and bondage for 70 years? It seems to me that this communist regime was one of the experiments periodically made by Nature, by God in order to show Humanity where one or another course of development leads.

In principle, the idea of Communism is wonderful: everyone is equal, everyone has equal rights and opportunities, according to one's needs and capabilities. This idea inflamed the minds of many people, some of the best representatives of Humanity on a worldwide scale. It is not said in vain: the one who wasn't carried away with the idea of Communism in one's youth has no heart; the one who is still enthusiastic about this idea in old age has no brains.

The main principle of the Soviet power was the principle of collectivism. We will discuss our conception of individual and collective development. In short, we can say that a human is the only being who unites extreme individualism and tremendously developed collectivism in his essence. This is a being, to whom an individual life is the most important, but he also belongs to the collective. And without collective movements, without collective life, there is no Humanity as an organized society. Communism is a model of society where the tendencies of an individual and individual activity are forcefully depressed, and all the essence of life is oriented on collective processes.

From the very first steps of the Soviet power; when Lenin, Trotsky and Kamenev united the labor masses, they placed the main emphasis on the collective. And these collective, organized masses were able to destroy the well equipped and trained Russian army; were able to sweep away the culture based on the thousand-year traditions. They managed to create a new type of society.

Stalin was a man of genius to put into practice the integration of people in a society with total shadowing and mutual control.

Much discussion in the West depicts the KGB as a dreadful force, controlling all Soviet society and entangling the whole world with its tentacles. This was really so. The KGB was always an extremely important part in the structure of the socialistic society. But considerably less attention is paid to the fact that the force of the KGB inside the Soviet Union was based on the each member being shadowed by his neighbor.

After destruction of Lenin's old team-mates and when the first tides of total repressions had been over, Stalin created the web-work system of everyone being shadowed. This started at the level of the Central Committee of the Communist party and extended to party organizations in the cities and regions. Then on to party nuclei at the plants, factories, at homes, thus out to every member of the society. It is necessary to remember that all parties except for the Communist party were prohibited. Concentration camps in Siberia were filled with people who were simply discussing the role of political parties in the Soviet Union drinking their evening glass of vodka.

The party nucleus was absolutely important, e.g. if at least 10 people gathered and only 2 of them were communists, the communists were to watch all the other people and report to higher authorities about any thoughts and movements. Collective activities were encouraged in this society: meetings, trips, team-work, joint activity, as well as communal life. Perhaps it is in Russia, in the Soviet Union that the principle of collective communistic life was introduced for the first time on such a mass scale. At the present moment such type of life is probably maintained only in the societies with tribal organization.

This system was incredibly and starkly embodied in the form of communal flats (shared apartments). But the truth was that these communes turned out to be quite far from the ideas of Campanella and other Utopians. A "communal flat" was a usual 2-3 room apartment with a bathroom and a kitchen, but every room was occupied by a family of 2-3 people. Primarily communal flats were created by way of "consolidation" of common apartments where a grandmother and a grandfather, families of their children, and families of their grandchildren lived. Another family was "temporarily" added to them and occupied one of the rooms. Afterwards the idea developed into communal flats in the form of huge hostels of 5, 10, and sometimes 20 rooms, a full family living in each room. There was a very long corridor in the flat and communal kitchen, where each family had a table, a range, a dresser, and later on a refrigerator. And the toilet was also one! For everybody. Sometimes, there were two. You can imagine what lines were formed due to that, as well as problems, conflicts, and stories!

Each family lived in its room, but it was impossible not to communicate with one another. Children were growing together, everybody knew who eats what, who wears what, who sleeps with whom, what relationships are in the families. Life of every individual was open and transparent.

The main idea of the proletarian leaders relied on the fact that in such living conditions it was not only impossible to make some kind of undercover political activity or transmit espionage data, but also impossible to earn some extra money, as it would become known right away. As soon as some of the neighbors got a new rag or a coat it was discussed by the whole commune at once. And if it happened too often, the communistic nucleus learned that straight away.

The system of shadowing ran through the whole country and was officially promoted as a national pride.

One of the main heroes of the Soviet pioneer movement was Pavlik Morozov. A small boy who gave out his father to the KGB, then his uncle and his mother. They were all shot. Pavlik was killed by the relatives who remained alive. Such was an ideal of a person for the Soviet power! The person who knew neither his origin, nor race, family, parents, and was devoted only to the party!

Such a system destroyed all discontented with the Soviet power in the 1930's, and exiled to the camps all the uncertain. Made the poor agrarian country one of the leading industrial countries in the world. Won the war, which cost colossal destruction and huge human losses. Created an absolutely new community of people and brought these people to total poverty and degradation.

The second important element of Stalin's ideal plan was the system of propaganda. From the very beginning of his political carrier Stalin started to win painters, musicians, writers over to the side of the Soviet power.

Young talented guys were encouraged and were treated kindly by the powerful.

They received better rations than other people. Special health centers were created for them. Those who had special party merits were granted state residences and cars. Everything in exchange for active labor for the benefit of the country of the Soviets, and for the glorification of the Soviet regime. The most outstanding creative figures were even allowed to travel abroad, although their families remained hostages for the authorities. And the people knew that if they made a false step, their relatives and close friends would suffer punishment. And the punishment was cruel and inevitable.

Soviet propaganda created a special image of the country of happy and cheerful people unaware of the problems and troubles of the capitalistic world. Foreign works were carefully selected before they reached the Soviet people. Chosen films and books depicted the life of Western countries in terrible colors. Recall the films of Kubrick and Coppola, books by Charles Dickens, Theodore Dreiser, and Jack London. Their works created an impression that Western life was full of snares and difficulties and that it was extremely hard to survive in the capitalistic world.

The "iron curtain" kept people inside the state in total ignorance regarding life in other countries. And in order to go to some capitalistic state it was necessary to rise to a very high level of the communist hierarchy. The wildest dream of the Soviet people was a trip to Eastern Germany or Czechoslovakia. And an opportunity to go to the socialistic (!) Cuba was absolutely fantastic. The Soviet people believed everything written in the newspapers. There were only five newspapers. The people believed in everything broadcast on the three TV channels, and lived a naive and happy life.

Therefore, in the years of my youth, when I graduated from the University with a "red" diploma (with honors) and was offered position at the research department, we couldn't even think to compare our life with the life of Western colleagues. Of course, there were dissidents. Of course, there were mass emigrations to Israel and the United States. But this was some other side of life. I recall when once with one of my friends we went to the park and he was telling me for a long time how it was possible to emigrate to the USA and how much better the life was there. Such talks were held only far from the walls, as in the Soviet Union all walls had "ears." Moreover, it was forbidden to discuss such topics in a group, it could have had quite a bad result. So, I remember that I was listening to his talk, very interesting, but absolutely abstract for me.

I had family, relatives who lived in Leningrad, interesting scientific work, and good perspectives. What was more, I was going in for alpinism, gave a lot of time and effort to it. The life was interesting and full. It opened exciting perspectives. And I had no grounds for discontent.

The scientific work I was doing was connected with absolutely physical questions: development of systems of direct transformation of thermal energy into electrical energy. An important topic still today. And at that time the prospects looked wonderful! It seemed that just a few more years of work and it would be possible to create a trans emission transformer. To transform the energy of the atom into electrical energy almost without loss, avoiding the need to heat water and spin turbines, as we do now.

Each day I came to my chair, turned on the instrument, and sat for long hours decoding the graphs, manually processing data, reading books and articles. Only some small moments darkened the unclouded horizon of my life. One of such moments was a collision with the KGB.

KGB

Every year, as soon as the summer began, we gathered rucksacks, said good-bye to our families and left for the mountains. After several years of practicing alpinism I managed to join a very good team which was performing on the level of championships in the Soviet Union. Each year with this team we were flying to the Caucasus, the Pamirs, the Tien Shan, and participating in mountaineering. We always went to places which were far from civilization, high into the mountains, where one could sometimes meet only shepherds, talking philosophically with stars and their sheep. Not in vain it is said that philosophy was born in herdsmen tribes. They had a lot of time to ponder over the meaning of life.

We were living in a small closed group for eight weeks, feeling our closeness with nature, mountains, and a little bit with one another. These were two months of persistent work, two months of free life far from everybody, far from the city fuss, far from civilization. Only we and nature. We and the mountains. The mountains were as if a model of the real world, where for a short period of time, for a few weeks, achievements were born, talents were discovered and ambitions collided. Where the play went on; but the play was on the level of life and death. Every time coming down, back to civilization we felt like people from another world. From the world of sparkling ice, fathomless azure, crystal air, from the world of trustworthy human relationships. This made us too vulnerable for some time to many small things of our usual life.

Once, after a particular trip, we came to the airport of Samarqand and encamped there. Samarqand in summer was very peculiar. There was a special life there: 400C heat (105 F), heaps of melons and various fruit. There were people in robes and turbans, fragrant tea at chaikhanas (Tea Houses) under the branchy palms near loquacious fountains. Alluring smells of shashliks and pilau. In brief, Asia.

It was very attractive to plunge into the Asian life, but we had to return to our home lives. As a rule, we never had tickets for the way back. In the mountains you never know how much time you will spend on one or another climb and what circumstances will make you descend. Therefore, we came to the airport in hopes of flying back by plane to Moscow, Leningrad or any other Russian city. And from there it was easy to return home.

That time the situation came out to be a deadlock. There were no tickets. The booking clerks were dozing off under air conditioners, exhausted, reacting to the beseeching looks of the passengers around, just like Queen Cleopatra to the smiles of merchants selling dates. The planes were full of organized tourists who had paid for the tickets beforehand, and we had no prospects to fly back to Russia. Having and a friend of mine, (in a couple of years he would get lost in an avalanche during a climb), decided to fly to Dushanbe, the capital of Tajikistan.

Dushanbe has never been a tourist city. Before the war it was a nice small town, but it never attracted tourists. Therefore, we were sure that we would be able to fly to our homeland easily from Dushanbe. We took the tickets, entered the cabin and occupied the places shown by the stewardess. Imagine: we were after two months spent in the mountain heights. Seedy ragged clothes, sewn in some places. As a rule, we had no women in the expeditions, we did everything ourselves. You can imagine how we were looking! Dark sunburned faces, tousled beards, ice-axes in hands - a full picture of wild men who had come down from the mountains. And looking like that we made ourselves comfortable in the cozy armchairs. It was a piece of civilization for us, and Coca-cola brought by the stewardess was a sip of life-giving moisture from another world.

The plane took off, and I heard English speech. I was actively studying English, but had no opportunity to practice speaking. Contacts with foreign people were extremely limited. Two people sitting in the armchairs in front of us were holding a map of Leningrad, and one of them was trying to explain to the other one where the Hermitage was situated. He was moving his finger around the districts which were quite far from the city center. I broke into the conversation and, using my modest knowledge of English, showed them where the Hermitage was. After that a simple slow conversation started. I was telling about mountaineering, adventures in the mountains, and institute life. A common, simple talk. The whole flight was only 45 minutes.

When the plane landed we said good-bye to one another and started toward the exit. The stewardess stopped me near the ladder,

"Sir, your ticket?"

"Here, please".

"And your passport, please".

"Here it is".

The stewardess quickly grabbed my passport and slipped it into her pocket.

"And what is wrong?" I asked in perplexity.

"You will have to wait, our colleagues will come now, and you will follow them".

These words brought me back to the reality, and I understood that I would have to deal with the KGB officers. Indeed, in a couple of minutes two men in mufti came up the ladder, they wore grey suits and had square faces with blank looks. One of them stood in front of me, another one from behind. And I was convoyed to the airport building before the eyes of the surprised Canadian guests.

I was brought into a big empty room; a table was standing under a huge portrait of Brezhnev. A man in mufti was sitting at the table, a real colonel.

"Take a seat," he pointed at the only chair standing near the table. I sat down on the edge of the chair. The colonel adjusted the lamp, or rather directed the light into my face, and asked,

"Well, tell me who you are and what you were talking about with the foreign agents? "

During the next three and a half hours I was playing dumb in every possible way:

described the amenities of our life in the mountains, demonstrated our medals and diplomas. I was trying to convince the colonel that our meeting with the foreigners in the plane had been absolutely fortuitous. He was looking at me not blinking an eye, without any expression on his face, sometimes asking questions. In three and a half hours he said,

"Very well, you can go." And gave back my passport.

"Comrade Colonel," I said, "Please, help us with the tickets. We need to fly home." He picked up the receiver, dialed the number, and ordered,

"Person called Korotkov will come to you now, make him a ticket..."

"Two tickets." I put in.

"Make him two tickets. To Leningrad. They should fly right now, the next flight".

An hour later we were entering the plane. My friend was greatly astonished as he had already run to the ticket office and had made sure that there had been no free places in Dushanbe, not a ticket.

A month passed. I was actively working and practically forgot about this episode. Of course, I didn't tell anyone about it. In the Soviet Union it was usual to hold such things inside.

And then, when I was working with my instrument in the laboratory, I was called to the telephone. A secretary of our rector was calling: "Konstantin, please go to the pro-rector right now." I went, and entered the pro-rector's room. Some person in mufti, with a square face was sitting at the table, and the pro-rector, having introduced me, went out backwards, bowing. Then followed another three and a half hour of talk during which I repeated all the history of my acquaintance with the foreigners word for word. He was clarifying which maps we had been studying. After which he said: "You can go. We will call you".

In several days I was called to the same room again. Another person was sitting there, but with the same square face. I repeated the same story again. It took another several hours. Finally, the person said,

"Konstantin, do you understand that your past behavior has drawn serious suspicion upon yourself. This can adversely affect your further life. Therefore, we will ask one favor of you. You work at the department of the institute, speak with people. Please, memorize what everybody is talking about. And report to us periodically. We are especially interested in the Jewish people."

"I am sorry, but I practically do not communicate with anybody. I stay at my lab and do only scientific work. I have no time for chit-chatting."

"Fine, think about this," the colonel said, "We will call you."

In a couple of days I was called to the same room again. There were two officers sitting there who started insistently offering me to gather information for them, describing what troubles I was to expect in case I refused to do it. I refused.

The situation repeated. After that I wasn't called any more. But that had direct consequences, indeed. In order to enjoy a career growth, to achieve the slightest professional success in the Soviet Union it was necessary to be a member of the

Communist party. In principle, any scientific employee after several years of work could have shown interest in joining the party, and he would have been admitted. Certainly, everybody understood that these were certain rules of the game: if you weren't a member of the communistic party, some career progress was practically prohibited. So, after some time the party organizer of our department came to me and said:

"Konstantin, are you going to join the party? "

"Of course," I replied, "Naturally, I was planning to."

"Well, then think, and we will discuss it in a short while with you".

I thought and decided that my life would change only for the better after that. After a time I came to the party organizer myself and said,

"Nickolay Petrovich, I would like to discuss an opportunity to join the communist party".

Having heard my words, he became slightly confused, looked for the flies on the ceiling for some time, and then said,

"You know, Konstantin, at the present moment we have no opportunities, and when we have I will inform you".

The "opportunities" were registers sent to every communist cell in the Soviet Union. They determined how many people and from which categories of employees - production workers, engineers, or scientists - were to be admitted to the party in one or another period of time. That question was repeated several times during the next months, while my colleagues were successfully joining the lines of the builders of communism. After that I realized that it was not just the absence of opportunities, but concrete orders concerning myself. And I gave up the idea of joining the Communist party for the rest of my life.

Now, when I am looking at all that from the perspective of my present situation, I see that maybe it was a will of fate which took me away from the false step, and enabled me to avoid the communistic mark in my biography.

There was also another situation. The head of our department offered me to go for probation to Canada. These were the 1980's. The system of scientific exchange between the scientists of the Soviet Union and Western countries was already working well. People traveled for one or two years to work and get knowledge in the other countries. That was supposed to be very useful, honorable, and gave colossal perspectives for further development. I agreed with pleasure, and started gathering the documents. I was successfully interviewed at various scientific counsels and societies. Everything went well, but suddenly at certain moment the procedure was suspended and my documents returned, and I was told that there were no vacancies. That was natural as my family name appeared in the "black registers" of the KGB.

INTUITION

Life went in its routine but interesting ways. Family, small children, work, trainings, trips to the rocks in spring and mountains in summer - the flow of life was stormy and rich. The mountains gave me first lessons of deep intuition. We had a painter in our team, Tolya, he was painting pictures and sculpturing. He was a wonderful person: deep thinker, philosopher, intuitive, and multi-talented. The reverses of fortune played a trick on him later.

He fell in love with a young girl, sixteen years old, having no family, no relatives. He took her to his home, and they started living together. Once police broke into his house at night, he was arrested and sent to the camp for three years for sexual seduction of the under-age. When I met him after three years, he was absolutely different person, had grown old, with a broken soul, - thus the camp life had changed him. He explained that the reason for his arrest was by no means his love affair with the young maiden, but his political views, which he expressed several times in public. That was typical of the Soviet power: to spy on each person, and if a slightest political hidden motive appeared, cruelly punish using any occasion.

We were in the mountains several times with Tolya. He gave me the first lessons of intuition,

"Look, Konstantin, mountains are a live organism. They have their own life, their own laws, and we shouldn't conquer them, we can only come here and live according to their rules, if we understand them. In order to survive here one should feel the breathing of the mountains, feel their state. What you need to live in mountains is intuition, in the first place"

We were coming up a glacier. The weather was wonderful. A beautiful peak was in front of us, our goal, the summit. We were full of strength and energy. The snow was creaking with the thin crust of ice under our feet at each step. We were in splendid mood. And suddenly Tolik said,

"Stop, guys!"

Everybody stopped.

"Well, let us rope ourselves together."

This means that two persons are tied to each end of the rope, hold their ends in hands, and go at the distance of 20 meters from one another.

"What for, Tolik?! There is a thin crust of ice on the snow - wonderful conditions!"

"No, guys, let us tie ourselves together. I don't like it."

We tied together and stretched along the whole glacier in long chains. I walked, looking around and breathing in the heady air of the mountains. I was in good spirits. And suddenly a powerful jerk knocked me off my feet. Some force pulled me along the glacier. The trainings had not been futile, I cut the ice-axe into the snow instantly, not even realizing that, using the outstretched legs in climbing irons, and started to brake without any idea to analyze what had happened. In a few moments

I managed to stop. When I raised my head, I saw that the rope was running away from me and disappeared in several meters, and my work-mate was not in sight. A flat snow field was stretching in front. Having fixed the ice-axe and the rope, I carefully crawled to the edge and saw what I had expected. A huge crack opened in the middle of the glacier, the crust of ice over the snow fell into it, and my friend was hanging at the height of 15 meters. The crack was covered with snow, and it was impossible to see it on the surface. When my friend had come into the middle of the thin snow bridge closing the crack, the snow hadn't been able to carry his weight and he had fallen through. Such things often take place in the mountains.

The rope had stopped his fall and had saved his life. In a few minutes he was back on the surface, and everything finished with little scratches and jokes. Tolya had felt the danger intuitively and had taken measures. Unless we had tied together with the ropes he could have died.

Such slopes in the mountains are very dangerous, as no one ever knows what is under the layer of snow and what cracks are waiting ahead. Only mountaineer's experience and intuition work here. I will give another example from the practice of mountaineering.

We were climbing in the mountains of the Caucasus; the weather was good and fair; by about 10 a.m. we were already standing at the summit. We were in cheerful spirits, the weather was wonderful: surrounding peaks were sparkling like crystal pyramids under the rays of a bright sun. Only in some places dark rocky bastions were beetling. It was late spring and all the slopes around were still covered with a thick layer of snow. Having admired the broad vistas, we started on the way down. The descent lay along a wide snow ridge, which brought us to the edge of a wall, where a long ice gully began and led to the lower glacier. We passed along the ridge and were ready to enter the gully, but suddenly an inexplicable desire appeared to me, and I exclaimed: "Guys, let us sit here, look around, and drink tea."

I had no idea why I said that. It was not much in my habit. We had a verbal understanding in our groups that during a climb we should not waste a single hour. We came up, looked around, - and went down to the camp, to get ready for the next rise. But there was still enough time, so after some discussions my idea was accepted. We settled on the ridge, took off our shoes, set the primus stove and started melting water for tea using pieces of ice broken off from the closest icy ledge. In an hour everyone was luxuriating in the sun, and no one wanted to move, although it was clear that we would still have to go.

Suddenly we heard some low sound in the atmosphere. That was a deep low tone. It seemed that it was everywhere and penetrated deeply into our very human essence. We all half-rose anxiously looking around. And we were suddenly tossed to the ground by an air wave. We were deafened from the crash. The gully which we would have gone down, collapsed as a single whole and went down. The snow cover broke in the very upper part of the gully, and, quickly growing stronger, a huge icy avalanche crashed down with thunder. We were watching from above how it rolled out for hundreds of meters onto the glacier. Just bare stones remained in the gully after it. Nobody said a word. It was clear that if we had not stopped to drink tea, we would have been in the middle of the gully at that moment.

Afterwards the intuitive feeling of mountains helped me several times not only to achieve success in alpinism, but also just to survive. And later on this feeling became the basis of my whole life and work.

Fig.46 New dreams, new ideas.

THE UNKNOWN

"There are two ways to be fooled. One is to believe in that which is not true; another way is to refuse to believe in what is true"
Soren Kierkegaard, (1813-1855)

My acquaintance with unconventional sciences began in late 70-ies. I worked at the department of physical electronics of Leningrad Polytechnic Institute and was also a postgraduate.

The topic of my work was deeply academic: "Change of work function of the polycrystalline wolfram under two-component adsorption of electropositive elements." Within a few years after graduation from the institute I built an experimental setup with quite good parameters for those times, published several articles and moved to the defense of dissertation confidently. I could await quiet work as a lecturer and leisurely scientific activity.

When we talk about scientists and research work, we fancy an image of some "pundit," sitting up all day long working with his instruments and from time to time making discoveries turning the world over. In practice really interesting and original ideas are the lot of few individuals, the vast majority of researchers do a boring routine work, obtaining the same curves hundreds of times, filling up dozens of pages with equations, and sitting all day long looking at the computer screen. At best this confirms an idea given by somebody earlier, at the worst - "a negative result is also a result." Only a few scientists get the Nobel Prize from hundreds of thousands. However, even this routine work requires great knowledge, mental effort, and gives rare moments of recovery, bringing deep satisfaction and a special kind of intellectual pleasure.

So, I was calmly working; one could have supposed that everything was wonderful from the outside. However as it often happens, "still waters run deep". A discontent was brewing in my soul. A calm routine work with the plan determined for future years did not suit my temperament well. And there was another circumstance, which created inner tension and discontent. Reading beautiful foreign magazines like "Surface Science" and "Physical Review," I always suffered from the gap between the Western scientific level and our Soviet level. Computer technologies and modern methods of data processing were at full speed in the West, any scientific center could order the most up-to-date equipment, but in our country we could only dream about that. I suspected that all the work for which I spent so much effort and time could be done in a considerably shorter time at a much higher level in the West. However these assumptions were only light clouds, which were spoiling the picture of my scientific activity for the time being.

But once I saw an announcement about a lecture: "What is the biofield?" The lecture was held by some lady, I recall neither her name, nor what she was. She was telling about quite simple things from my present viewpoint, - about telepathy, extrasensory individuals, and the Kirlian effect. But this lecture made a great impression upon me: it turned out that there was still a real Mystery in the world, and there were scientists who studied it. "I would love to be one of them!" I thought with a deep-seated grudge. I even started seeking who had done all this and where, but having learned nothing, I put this thought far back into my memory.

But then, my scientific adviser, senior lecturer Norbert G. Ban'kovsky, called me and asked,

"Konstantin, have you heard something about the Kirlian effect?"

Here I suppose it is time to tell the prehistory of this question. In the beginning of 1970's, the Kirlian effect was studied all over the world ... except for the Soviet Union. International conferences were held, books were published, but little was known in the scientific circles of the Soviet Academy of Science. Such state of affairs caused anxiety in the Soviet Committee of Science and Technologies - the head organization, responsible to the Central Committee of the Communist Party for the development of science in the USSR. The Committee instructed the Presidium of the Academy of Science of the USSR to investigate the question of the Kirlian effect. The Presidium, having studied the problem, sent it for detailed examination to the Council on physical electronics of the Academy of Science of the USSR. The Chairman of the Council, academician N. D. Devyatkov requested his good friend, Professor N. N. Petrov, head of the department of physical electronics of Leningrad Polytechnical Institute, to review literature on this topic and report at one of the sessions. Professor Petrov made the same request to his subordinate and friend - senior lecturer of the department Dr. N. Ban'kovsky, the latter called a junior scientist, Konstantin Korotkov. This was the first call of my Fate, which sharply changed not only my scientific career, but also, my whole life.

"Konstantin, have you ever heard about the Kirlian effect?"

And this was the beginning of my journey into the world of the enigmatic fluorescence, the world of subtle energies, of the magicians wearing ties, and serious scientists with anxious hearts.

"Yes, Dr. Ban'kovsky, just a week ago I listened to a lecture on the biofield, and this phenomenon was mentioned there. As far as I understood, this is a method of investigating the biological fields of humans and plants."

"Absolutely correct, something like that. Our department was commissioned to find out what is really scientific in this effect; too much loose talk and speculation around this phenomenon - such a position always causes distrust. What do you think about reviewing the literature on this topic and continuing the session of the Committee of physical electronics, of course, parallel with your main work?"

"I agree, how much time do I have for this work?"

You should understand, I agreed straightaway. Only a short time before I had been unsuccessfully trying to find out who had been studying unconventional problems and where they were, and then a real opportunity to do it myself appeared. Dr. Ban'kovsky informed me that the report would be in a month, and I was promised that literature would be sent from the Council.

"However, Konstantin, in order not to waste time, go and search in the library something on this topic."

The next day I went to the State Public Library. I enjoy working in this library. It is located in the very center of St. Petersburg, in a beautiful ancient building with wide staircases and high ceilings. Solemnity still reigned in the hall for scientific work, with high lancet windows which opened on the park with the monument to Empress Katherine, and the rustle of pages ceased to be heard in the multi-story height of the hall. Having looked through the catalogues and having studied the "Citation Index," I found several articles in rare American magazines and the only copy of a book "Kirlian Aura" edited by Stanley Krippner. That book was in Moscow, in Lenin's Library - the central and biggest library of the Soviet Union. The next day I was on a train to Moscow, and at 10 o'clock in the morning I was holding the longed-for book in my hands.

The report at the meeting was a success. I recall that I was terribly excited and, therefore, I was reading the text of my report from paper, keeping my eyes glued to it. However, when the questions on the merits of the case started I forgot my anxiety and, as it was found later, had replied quite clearly to all. After the meeting academician N. Devyatkov called the three of us - Petrov, Ban'kovsky, and me, thanked us for the work done, shook my hand, and suggested we should study this problem in more detail at the department. Naturally, the laboratory of Dr. Ban'kovsky took that up.

Could I have then thought, starting this work, that this would be the road to new knowledge, that it would open new horizons of human existence, and enable me to join the international scientific stage?!

Clouds wreathing under the window of the plane. Somewhere under them is the Atlantic ocean. And suddenly the sun creates a fantastic picture, - the shadow of the plane falls onto the clouds and a rainbow lights up around this shadow. Looking through the window, I see this shadow, see this rainbow around it, and the deepest meaning opens to me in this glowing aureole. Only once before had I had a chance to experience such a feeling of cosmic consciousness.

Together with a friend, we were ascending a steep wall of the Dombay mountain in the Caucasus. It was our last climb in the season, the final one. We were full of strength and energy. The wall was the most difficult in the surrounding mountain region, but we had already passed the most serious part by the middle of the day. So, after another plumb rope we climbed to a ledge which was enough not only for setting a tent, but also for lying down, and we decided to stop there unhesitatingly. We started cleaning the stones, using the ice-axes as spades and turning the inclined shelf into a horizontal plane. Suddenly Peter exclaimed: "Konstantin, look!" I raised my head, looked around and stood motionless in admiration.

We were standing on a small ledge of a steep wall, going down for half a kilometer. In reality we were hanging on a narrow ledge of rock midway between sky and Heaven. No bottom of the wall was visible. A shroud of clouds was under us. They were so dense that they seemed to be solid. We had an impression that if we had jumped onto these clouds, they would have embraced us, and it would have been possible to jump and swing on them, like on a layer of dense cotton. And these firm clouds stretched far away, until the very horizon. The majestic peaks were rising around them, most of the summits were lower than our level. Small clouds were flowing above the firm ones and looked like air angels. It seemed that we were sitting on the doorstep of Paradise. Looking at the angels' huts from above, watching the calm flying of the heavenly angels. And all this was shimmering in the rays of sunset, changing colors from golden to turquoise. Then reddish tints started to appear. Then the sun itself fell under the layer of clouds. And suddenly everything around got illuminated with a pink color. This was the color of the divine glow.

We were sitting, motionless, looking at that beauty, having no idea that the next day would be the day of fatal trials, that we would be several times balancing between Life and Death, as if we were some figures participating in the Universal Game between the Evil Demons and the Good Angels. And the broken leg and stone-bitten bodies would be an easy payment for the participation in this Game.

I AND WE

So, what is it that determines the Line of Fate, the train of random, or as they seem to us, random events? Why one person is kept, taken care of and led from one situation to another one by his destiny, and the other is watched and punished. And is he punished? Maybe it is we who arrange what is happening to us, and the outside forces just give us conditions to realize that.

Not long ago, standing at a crossroads, I saw a "Zhiguli" car passing by. A mumper with a heavy look was sitting at the wheel next to a pretty woman. I was surprised to recognize my former friend. Some time ago we were very close. I have never met a more talented person. Early in life he was one of the wittiest, brightest, and most brilliant young men. He was the soul of any company, a women's favorite, faithful friend, cheery fellow, a person with such a special sense of humor that he turned any banal thing into a work of art. One can not learn to do this; it's a gift from birth. Charley Chaplin and Mark Twain were among such people.

Unfortunately, the majority of us are quite boring, our thoughts are trivial, they follow their usual road limited by certain borders, and we see things from their trivial side. But particular people are gifted with a special vision of the world, perceiving it originally. They open something different in every event, as compared to us. Inimitable story-tellers are born among such people, gathering the cream of society around them. The brilliant salons of Paris, London, and Madrid were mostly based on such individuals who were able to create a special atmosphere around them. And stories about them often surpass their own creative work. Therefore, as a rule, these individuals create only in the presence of an outside impulse. They always need somebody who would excite their imagination, wake up their inspiration; then they can start dreaming and gush forth: be effervescent, create a firework of topics and images. But as soon as they stay alone and go deep in thought inside themselves, the process stops. Sometimes they can not even reproduce what they were brilliantly improvising an hour ago in the company of beautiful ladies. Oscar Wilde was one of such persons. We know that his creative life was much richer and diverse than the works of art written by him. And how many other great geniuses of impromptu remained in the memories of friends and visitors of literature salons!

My friend was such a person. But there was a feature which surprised me greatly for many years. This feature was his bad luck. It seemed that the most diverse problems and misfortunes were falling down onto his head. He always got into some scrapes. He went to the mountains, and the weather turned bad there, catastrophes took place, people died. He was doing scientific work, and after a while his team broke apart, inexplicable squabbles and petty troubles started, the work stopped and his own colleagues took away all the materials. He started doing business, and after some time brought to bankruptcy not only himself, but also his partners who had been rich and lucky before. It seemed that those misfortunes followed one after another having absolutely no connection with or dependence on his brilliant character and outstanding wit.

But later, when I was able to look at all this from another viewpoint, of the development of the human spirit, the human soul, I suddenly realized that everything that had taken place had a deep inner meaning. Everything happened only because of the fact that this person had always been aiming at himself. He had always been concentrated on himself, and nothing but himself, his own claims and success, had ever deeply interested him.

All his life, all his effervescent merry-making, and all the firework of his wittiness needed only admiration from other people. All his work was done in order to demonstrate how wonderful and great he was. He needed no money as such, this wasn't the main criterion of his work. He could easily turn from the planned way, leave everybody and go aside, if he thought that more popularity was waiting for him there. He valued neither friends, nor the beloved. Ambition was his main driving force. He needed people who admired him, who looked at him up. His Ego had always been the center of the Universe. Although he hid that in every possible way, and never demonstrated in public (he was too "wise" for that).

Apparently, the essence of such approach, even concealed from superficial view, was manifested in profound laws of life, and the life itself at its basic level gave him what he deserved: led to collapse and catastrophes. Other people who were involved in the sphere of his life and activity were often suffering in those catastrophes.

With time every person undergoes transitions from the outside to the inside world. The speed of one's inner time changes. The orientation of the whole life circle changes.

A child fully depends on people surrounding him. An infant can not become a person without them. They provide love, care, understanding and nourishment. And this is not only physical food. There are many examples of children brought up by animals: monkeys, wolves, goats. When they got into the human society they were not people - those were small monkeys, wolves, and goats. All attempts to turn them into people failed. Information a child or growing teenager receives from others and the outer world is the most important for his/her development. Childhood and youth is the time of communication, time of friends, and time of perceiving the world through other people.

With the course of time a person plunges into his own troubles more and more. He concentrates on his only world, and gradually other people surrounding him stop playing the important role. And, finally, alone he passes to the final meeting. Nobody plays any role in this meeting, nobody exerts influence.

So, what is the ratio of individual and collective in a human being? Is there a collective consciousness, which was once discussed by the great psychiatrist-intuitivist Jung for the first time and which is being increasingly used in modern conceptions? What is it? Is it just an ideal notion or is there some physical reality behind it? Is this notion material or ideal? And what is more important in our life: material or ideal?

Looking at the chain of events of my own life, studying the history of Humanity, even distorted, known to us from chronicles, more and more often I come to the idea that all events, all historical phenomena represent material embodiment of the spiral of development of the human spirit. The history is run not by material stimuli, not by ambitions of particular governors, and not by religious confessions. It is moved by spiritual impulses. And as in many aspects of our life, these impulses represent a non-linear sum of inner tensions and outer forces. They are formed from stresses and conflicts, which have always existed in human society, and from the influence of external, natural and cosmic factors. The history is guided by individual and collective spiritual impulses in many respects.

In order to make this thought clearer and show it both in the development of individual lives and in collective processes, let us first dwell on a general model of the human being, which we have been following in recent years. The human appears in this model as a unity of three principles, an indivisible and inseparable unity.

One of the representations is physical. This is the physical body, physical processes, -everything which unites us with the material world, the world of animals, from which this body originated and to which it belongs in many respects. Our potencies, desires, connected with the satisfaction of material demands and sexual needs - all these belong to the material body. Nature gave birth to it; it belongs to nature, and will go back to the world of nature after its owner's death.

The next part of the indivisible substance of the human being is the spiritual substance. This is the soul, the informational field; this is the ideal, nonmaterial part which exists in unity with the physical body, but which can come apart from it and exist in the form of separate substance. There are many descriptions of separation of the soul from the body during clinical death, when an individual could see his own body and doctors bustling around him from the outside. In the USA I had a chance to visit the Monroe Institute, whose founder had got from his youth an ability to "travel" out of body, move in space and in time, visiting friends and relatives. He founded an institute and started successfully teaching all interested people to "travel" out of body. So, look around, maybe in this very moment you are surrounded by spirits flying in the air and watching your behavior with interest. A joke. But who knows? Our soul is the carrier of our feelings and emotions. This is all what accumulates ideal aspirations, feelings, and desires, unconnected with the direct maintenance of life processes.

The Soul is the connection of the physical body with the spirit - the third component of substance. The spirit, embodied in the world essence, cosmic substance, in one God for the whole Mankind, however having many (although denumerable number) modalities.

And these three substances: the material body, soul and spirit make up an individual human being. If one of them is lost, the fullness of life is broken. When a person looses his soul he becomes an animal. When an individual loses connection with the spirit, he turns into a machine. It doesn't mean though that a person should necessarily be religious and visit all church services. There is a great amount of spiritual atheists and soulless churchmen. Connection with the spirit is first of all the orientation of a human on high ideals, serving the aims which are beyond mere life support. Even if the latter requires the most part of one's time. But if once a person can go to the balcony, feast his eyes on the hidden sun, and freeze in admiration at the beauty of the Divine creation, his soul absorbs a particle of the universal Spirit.

The key-note we elaborated implies that the development of both the individual person and Mankind as a whole is first of all the development of the soul, individual spiritual needs and their satisfaction. All human life is aimed at satisfaction of needs of the three substances: physical body, soul and spirit.

It is impossible to live without normal conditions for the material body. And if a person is hungry, he will first of all be thinking about feeding his material body. If he is sexually preoccupied, the hormones flowing in his body will subconsciously determine all his thoughts and acts. This is the first basic level, and one needs to satisfy it in order to give life to all the others.

The first thousands of years of Human existence mainly passed in search for the best ways of satisfying this basic level. But gradually, getting more and more products, satisfying physical needs, all attention began to be switched to the level of soul. People started looking for ways to give pleasure to their soul. This created the basis for the power of church. Gave spur to the birth of music and arts. This led to great importance of beauty in human life, and people were ready to pay more and more for beauty, a sweet song, and a nice view from the window. All this talk that jewelry, fine clothes and articles serve for sexual attraction of one another is materialistic nonsense. People have always made love - this is one of the most important stimulus of life, and both jewelry and attractive clothes are not so important. Rings on fingers and in the ears, first of all, please their owners and show their social status. People become attracted to beautiful toys at any age, and adults' toys as compared with children's are much more expensive. We have a longing for things and events which please the soul. And the more an individual soul develops within centuries, the more exquisite and refined joys and pleasures it requires.

Looking at history we will see that it develops progressively, regularly repeating the past at another higher level - spirally, and each next wind gives better and better methods for satisfaction of needs of the human soul. The history of Mankind is the history of human spirit.

An outstanding Soviet historian Lev Gumilev developed his concept of history. He asked himself a question: why people living a quiet life, were suddenly rising in one gust, were forming invincible armies and were conquering huge spaces, creating states. And after the lapse of several centuries they dispersed in space and time, leaving memory only in chronicles and legends. The Vandals and Goths, Genghis Khan and Timur, the unconquerable Ottoman knights - they left just memories. Why some states dominated for a time and then passed into nonexistence again? Lev Gumilev developed a concept which explained these processes. His idea is based on the fact that at some moment a certain outer impulse gives extraordinary activity and powerful inspiration to peoples living on a certain territory. This impulse gives birth to people's inner energy, "passionness" as called by Gumilev. Passionness is the energy of the soul creating greatest incitement to individual development and individual aspirations. Passionate people can't be calm. They can't live a monotonous ordinary life, they need journeys, activity, performances. The approval of their companions is much more important for them than their individual wealth or individual achievements. They do not care about their own life, their own prosperity and well-being. Spiritual idea - this is the moving force of these people. Thus it was in the times of Genghis Khan, who was able to create an innumerable army, raising it with the spiritual idea of fight against Chinese oppressors. Thus it was during the first Christians surviving extreme pursuits and sufferings for the sake of spiritual concept. It was in all great wars, when armies fought driven by spiritual idea. This is the course of history.

What is more, these ideas haven't always led to humanistic manifestations. Wasn't it the spiritual idea that drove the troops of the violent Hitler? The idea of creation of a new race, new community of people, true Aryans, the only descendants of the great tribe. Wasn't it the spiritual idea which enabled Russian people together with their allies to win a victory in this war, to overcome the trained, perfectly equipped fascist army? And wasn't it the spiritual idea which helped people to survive in the inconceivable conditions of concentration camps: the fascist, Soviet, Korean, and Chinese?

The history of life of my father has always been a bright example for me. A difficult life. Life inspired with great and spiritual idea. He didn't live very long - 70 years, and passed away during a distant business trip from inner overstress. In the end of life he wrote memoirs. These pages written by his hand are a part of the history of the Twentieth century. A part of those invaluable crumbs of life, which disappear, often leaving no traces. And new generations, unaware of these crumbs, having no idea of this experience, repeat the same mistakes.

Mankind should not forget what it has gone through, what took place in its spiritual history. Therefore, I quote these memoirs in full here, the way they were written, with all stylistic peculiarities, as an invaluable historical document.

All the life of my father passed under the Soviet power, and all his life he was a devoted Soviet communist. When I asked him,

"How could you survive the camps, you suffered so much?! You suffered humiliation and persecution for many years. And after that you can say that the communistic idea is true?"

He would always reply,

"Yes. Because these were particular mistakes of individual people, and the very communistic idea is the truest!!!"

Looking back from the Twenty-first century, we can compare political and economical development in Eastern and Western Germany, South and North Korea,

Cuba and Mexico. We can see now that the situation was not specified by a particular good or bad leader. The fact was in the idea itself! The idea which was totally fallacious. The idea which put the collective over the personal, which enabled the collective to suppress the individual, to usurp it, to put into the limits of an ant's order, and, having bringing it to the high degree of perfection, led to the full ruination.

The Twentieth century is the century of historical lessons. It is as if history intentionally made experiments, demonstrating to Humanity what is good and what is bad. A democratic society is the society of individual opportunities and collective control. And a fascist-communistic society is the society of individual suppression and total control. Is it possible to keep in mind these lessons and never repeat such experiments?!

But what a power was in this spiritual idea for people who went through the Stalin camps if they could stay true communists at that, and not for a second had they any doubts in the truth of the course chosen by the party and government! The level of "passionness" of these people was very high.

Judging by everything, the beginning of the Twentieth century was the time of powerful passionate influences all over the European world. A calm period came in the middle of the Twentieth century, when the processes taking place in the world had a sluggish, lengthy character. And people with heightened passionness, heightened energy were mostly directed to individual performances. Therefore, it was the time of scientific discoveries, projects, time of writers, painters, and armchair scientists. By the end of the century the intensiveness of processes started growing again. The first impulses began in the 1970's, they caused the strain of the 1980's and explosion of the 1990's. The energy of the Universe began growing rapidly, increasing the passionness in sensitive people and sharpening chronic diseases. Humanity came close to the new phase of its development. But let us discuss this all step by step.

So, before moving further in the story - to the 1970's, we should come back to the middle of the previous century. To the life lived by the married couple, the Kirlians. To the atmosphere reigning in the gathering forces of the Soviet Union. The farther this time moves from us, the more difficult it is to imagine with one's own eyes, to understand how people could survive all these hardships and what were the People from the Capital Letter. The Kirlians belonged to the generation brought up on the ideas of the World Revolution, the ideas of Stalin, the generation believing in the light of the next day and ready to sacrifice today for their present for the future. Personal acknowledgement was not important for them; serving the Cause, the Mother country, and the Party was only important. Such people were the married couple, the Kirlians, the same kind of person was my father, whose memoirs I found only after his death in 1978. These are live words of the epoch, written by one of its common builders.

MEMOIRS OF FATHER

ABOUT THE MEETINGS WITH A GREAT PERSON - FATHER OF SOVIET AVIATION - ANDREY N. TUPOLEV AND HIS WONDERFUL ADVICES AND RECOMMENDATIONS.
GEORGY K. KOROTKOV

Beginning of my working life

Year 1927. This year is wonderful to me because being in Rostov-upon-the-Don I finished an evening school. At the same time I was working as a mechanic in a locomotive depot of the North Caucasus railway during the day time; there I started my official labor activity as an apprentice of mechanic in 1923 being 15 years old. Then I came to Moscow to study in Moscow Institute of Transport Engineers. I was greatly interested in aviation, and basing on my knowledge those days I supposed that only the steam locomotive and flying vehicle were the most perfect technical achievements of the beginning of the XX century. I had heard already then that in Russia there were talented creators of flying vehicles - aeroplanes, and that according to a decree of the Soviet Government, signed by Vladimir I. Lenin in 1918, Central Aerohydrodynamic Institute was founded. An outstanding well-known aerodynamist Prof. Nickolay E. Zhukovsky was appointed the head of the institute, and a talented engineer Andrey N. Tupolev was appointed his deputy.

At the end of 1927 The Moscow Party and Moscow Komsomol Committees mobilized several students, party members from Moscow Institute of Transport Engineers into aviation industry for a short period in order to increase the party and Komsomol layer. I was one of the mobilized (Georgy K. Korotkov, member of L.Y.C.L.S.U. (Leninist Young Communist League of the Soviet Union) from 1922 and member of Party from 1927; born in 1908). I was sent through the Krasnopresnensky district party committee to the aircraft construction plant N 22, based in Phili, where a German concession Junkers and a big underground group of Trotsky followers were then situated. Concession Junkers produced one-engine two-seater airplanes. The concession had been founded by V. Lenin as a temporary measure for the creation of the Soviet aviation industry, since at that time Junkers had had the best samples of airplanes as compared to the other countries.

Firm Junkers brought equipment, materials, professional craftsmen and workers (comment: according to the Versaille Agreement, it was forbidden to produce airplanes and other military defense products in Germany). V. Lenin had then been convincingly proving that all the best foreign achievements in engineering should have been used in the country of Soviets.

The Concession Junkers played a significant positive role in the creation of the first stage of the Soviet aviation industry with active assistance of Prof. N. Zhukovsky and his deputy A. Tupolev.

At the end of 1927 Andrey N. Tupolev with a group of directors of the Main aircraft administration and plant N 22 visited the plant and studied the equipment and personnel thoroughly. The purpose was to investigate the process of aircraft manufacturing, so that the Soviet designers could produce airplanes which would be much better than those manufactured by Junkers, as it was planned to cancel agreement with them.

In 1928 the agreement with Junkers was annulled. Production of Junkers aircrafts was coming to the end and at the same time manufacturing of domestic airplanes with metal-corrugated protection (hauberk-aluminium) was being organized: twin-engine bombers designed by A. Tupolev, type "TB-1", and float twin-engine hydroplanes - reconnaissance planes, type "R-6", also designed by Tupolev. Then the plant started organizing production of three-engine passenger metal airplanes.

The plant rapidly developed receiving big government orders, but at the same time there was a great lack of engineering-technical personnel, skilled masters and workers. Andrey Tupolev time and again showed active care in bringing up personnel; he periodically visited meetings where the secretaries of party cells were sometimes invited, including myself (in 1928 I was elected secretary of party cell of mechanical welding workshop. In 1928 at one of such meetings Andrey Tupolev made a request to all, especially to young people, about filling up the lacking professional personnel by way of technical study by correspondence.

The administration and party organization of the plant took all those recommendations into account and bravely proposed gifted workers, who later on became heads of large departments. Circles for the preparation for Higher Institutions with teachers from engineers were organized. It's worth mentioning that there were big problems with urban transport, as a matter of fact there was no transport in Phili then. In order to get to work we took a tram (from the center of Moscow) to the Byelorussian railway station, from where we took suburban train to station Phili and then went on foot.

In autumn 1929 comrade Ledus was elected first secretary (on administrative work) of the party committee and I became second secretary (on agitation and propaganda). The collective of the plant was proudly building first progressive airplanes of A. Tupolev. In 1930 I started studying at Moscow Aviation Institute named after S. Ordzhonikidze as a full-time student at the airplane-production faculty. I graduated in 1935. During the period of studies I kept connections with plant N 22, lived in a new village "Phili" near the plant, during my vacations I worked as assistant to the head of workshop, technologist on general assembling of airplanes, and then designer at SDO Special Design Office). It was very difficult to attend Tupolev's lecture at Moscow Aviation Institute, as the hall was overfull, and the lectures were very interesting and important; they were mainly overviews; lectures for senior students were about novelties in foreign countries.

Aircraft designer

I had a chance to work as chief designer. We were surprised and glad that despite of great workload with various types of new planes ("R-6"; "TB-3"; two-boat six-engine hydroplane; land "Maxim Gorky"; "SB"; one-engine ATN15 and four-engine plane specially for the flights in the Arctic Zone and flights to the USA through the Arctic Zone, and other types), Andrey Tupolev came to the plant, gave advice and assisted. He had a good memory, remembering "old" professional personnel by faces and names; he joked in discussions, and was always interested in new engineers. According to his order to the designers, not just a good sketch of assembly of the plane should be created, but the design and calculation should be thoroughly studied from all viewpoints, such as: strength and weight, technology of production, safe maintenance and repairs or exchange. Solving these problems for every plane would enable our Soviet country to produce the best airplanes in the world, which fly faster, farther and higher than all others.

Owing to great efforts and concern of Andrey Tupolev as early as in the first half of 1930ies many professional personnel and production workers were gathered at large aviation plants, and our Soviet Fatherland started to produce)the best airplanes in the world.

Beginning of repressions

Our happiness due to the success of our aviation, the Soviet aviation industry, and other fields of national economy of our Soviet Fatherland was darkened by the fact that the best creative forces of our aviation were arrested as a result of subversive, provocative activity of imperialistic, conservative, and fascist circles. A. Tupolev, V. Petlyakov, V. Myasizhev, N. Nekrasov, A. Cheremuhin, I. Prozenko, Ozerov, S. Korolev, N. Bazenkov, and many other specialists and big patriots of the Soviet Homeland were arrested. It was a strong blow and misfortune to our whole Soviet country, inexpressible in words.

The arrests of skilled comrades at plant N 22 and other aviation plants, as well as in other industries of the national economy of our Soviet Fatherland, first and foremost defense industries, caused by provocations directed to detriment of our Soviet country, were wave on wave following one after another. The remaining specialists were working day and night with great zeal in order to fill up the comrades who had left. The undeserved mass arrests began from the end of 1936 and continued for a very long time in 1937 and 1938, and considerably decreased in 1939. This adversely affected the performance of plants, design and research organizations.

We did not believe and did not surmise that such great Soviet patriots and world specialists as Andrey N. Tupolev and his associates became public enemies, and we hoped that this mistake of the state scale would be clarified in the nearest time and everybody would be rehabilitated. I and comrades Volkov, Shpak, Vasserman and other comrades expressed this opinion in the party organs. Of course, certain circles did not like that, as the initiators of the undeserved repressions considered that the lack of faith in the activity of the People's Commissariat of Internal Affairs (P.C.I.A.) was thus created. It was necessary to show special firmness of Soviet patriotism from the side of the wrongly repressed in order to take all possible measures for the consolidation of power of our Soviet State in practice, in the situation of distrust, to their personal detriment.

The great immeasurable merit of Soviet Patriotism of Andrey N. Tupolev consisted in the fact that under the conditions of cruel repression he proved to the highest leaders of the Party and Soviet Government (he wrote a letter addressed to comrade Stalin in the beginning of 1939) that all grave accusations were, provocative, hostile, aimed at undermining our Soviet home front, particularly Soviet heavy aviation. He expressed that the best examination of the Soviet patriotism of specialists - prisoners was to use them in their direct professional work, and the results of their work would characterize their Soviet Patriotism. Andrey N. Tupolev contracted big liability to create the best dive-bomber in the world, and this engagement was met with great honor before the start of the Patriotic War 1941 - 1945.

Thus, many people of different fields of industry knew A. Tupolev, spoke highly of the father of Soviet aviation and expressed great astonishment and regret at the repression. They deeply believed in the innocence of Andrey N. Tupolev and maintained that repressions of such well-known specialists and Soviet patriots as A. Tupolev only gave good to the enemies of the Soviet State. Unfortunately, at this moment the investigating agencies of the People's Commissariat of Internal Affairs, mainly in 1937 and 1938 played into the hands of enemies of the Soviet State and this continued to a lesser degree until 1954 (until Beria with his enemy gang was done away with under the new leadership of the Party and government of our Soviet Fatherland).

I and many other comrades working at the plant N 22, Moscow Aviation institute, in the people's commissariat, and in aviation industry were expecting that the great state mistake would be corrected and A. Tupolev would start active work with the collective. Unfortunately, this did not take place in 1938, and, thus, the Soviet Homeland suffered great loss.

I emphasize again that owing to the convincing requests of Andrey N. Tupolev, from March 1939 the imprisoned specialists were used in according to their profession regardless of the article of indictment or conviction, which led to great positive result in the consolidation of the defense of our Soviet Fatherland. By the beginning of the Great Patriotic War of 1941-45 quite modern, best samples of military equipment were created in all spheres of defense: aircraft and tank design, artillery, submarines, black and non-ferrous ore mining and processing, metallurgical, coal-mining and building industries, as well as agricultural, food industry, etc. That was done when the specialists convicted by mistake and showing high Soviet patriotism in morally difficult conditions of imprisonment were rehabilitated. This important historical fact is little known at present time and, what is more, not many people know about the great positive role of A. Tupolev in this matter.

It is difficult to imagine that historical time interval (1937-39), which was severe for our Party and our Soviet people. It was filled with fully counterfeited political indictments and provocations, organized by reactionary circles of imperialism against the best Soviet patriots with the purpose to undermine and temporarily weaken the home front of the Soviet Country and then launch a war and destroy our country of socialism. Great force of Soviet Patriotism and colossal Soviet optimism of Andrey N. Tupolev was needed in order to make wise conclusions in such situation and put into practice the measures directed to the consolidation of defensive capacity of Soviet Fatherland producing best samples of airplanes in the world.

Arrest

I was arrested at night, at 3 a.m. on November 6, 1938. It was after a grand meeting dedicated to the October Socialistic Revolution held in Phili, in the club of plant N 22 (where I worked as head of department of the Bureau of Rationalizers and Inventors from February 1938). I came back home very late and tired, but satisfied with the fact that the plant went ahead of schedule by the results of production indices. Therefore, the people's commissar of aviation industry comrade Michail M. Koganovich thanked all the collective of plant N 22, and many comrades were put forward for a governmental award including myself. (The Order of the Red Banner of Labor).

Before the October Anniversary of the Socialistic Revolution in 1938 the mass arrests in aviation and other lines of defense industry were repeated. Regardless of the fact that Ezhov (people's commissar of the People's Commissariat of Internal Affairs (P.C.I.A.) of the USSR and State security) was ousted and convicted, and L.Beria was appointed instead of Ezhov, repressions in accordance with counterfeited materials and slanders continued, pleasing enemies of Soviet People. After Ezhov, as told by eyewitnesses, four categories of the convicted were determined using the corresponding physical methods of investigation.

After being arrested at home I was brought to the square of Dzerjinsky into the basement of KGB building, where with many comrades we were kept and then were brought to Taganskaya jail in a cell overcrowded with those who were under investigation, about 70 people. But there was no one from the aircraft plant in that cell. Comrade Melnikov, under investigation from 1937, member of the Bolshevik's party from 1912, chief engineer of Kolomensky steam-engine plant, was in that cell. During the October holidays he informed me about the situation. Comrade Melnikov fell ill (liver and other diseases) and spent approximately 10 months in Taganskaya hospital, where he had been transported after he collapsed due to inhuman methods used during his interrogation.

I was taking exams until April 23, 1939, and within six months I became certain of mean methods of falsification, slander, and intimidation with assault that were used in order to wrongly accuse and force to sign a fabricated indictment. What was more, the investigators were without scruple declaring that no one would be released from the jail walls, as that would have undermined the authority of P.C.I.A. I stood all the attacks of the examination and refused to fabricate accusations both against myself and the others. The Military Office of Public Prosecutor rejected to take my case and convict me at the military tribunal, as there was neither any evidence to that, nor my "admission" in a counter-revolutionary activity. (At that time at the urgent request of chief military public prosecutor Vishinsky, in order to convict somebody it was enough to present admission of the accused without any confirmation with the facts and witnesses. This law, causing full violation of Lenin's rules and international jurisprudence was repealed after objective and fair investigation, after the liquidation of Beria). After a might-have-been accusation I was brought to deputy people's commissar of P.C.I.A. of the USSR Kravchenko, who was trying to persuade me to put my signature under an anti-Soviet anecdote of any contents, which I should then "recall" and which did not really exist. But after signing that anecdote I would have been convicted for 3 or 5 years, and what was more, I would have been sent to my professional work as a designer of new airplanes at a special object together with great specialists in aviation, imprisoned A. Tupolev, V. Petlyakov, V. Myasizhev, N. Basenkov, D. Markov, Ozerov, A. Cheremuhin and other comrades, who had deserved authority and respect by their work before the repression. Deputy people's commissar of P.C.I.A. Kravchenko declared that specially good conditions were created for the imprisoned professionals at the special object, which enabled to implement State orders in aircraft construction in a short period of time, and that P.C.I.A. organs would assist in that.

When I started explaining deputy people's commissar Kravchenko that I hadn't liked and didn't like anti-Soviet anecdotes and thus I had never joked like that, that I had been brought up in Komsomol spirit from 1922 (from 15 years) and had been in Lenin's party from 1927 (from 19 years). I told that being a member of a party bureau and propagandist I had to organize an active fight against falsifications and misinterpretations of the policy of party of Lenin and the Soviets. I told that I would not sign any false evidences neither against myself nor the others, whatever would have been done to me, that I would not mislead organs of the Party and Soviet power, my friends and relatives, but I stressed that I was ready to work in my profession 12 hours a day. Deputy people's commissar Kravchenko replied to this: "It's a pity, you'll be lost holding a spade in the camps, think better one-two days, and it's necessary to sign the evidence, think..." After those words deputy people's commissar stood up and went out, leaving me alone in the investigation room.

My consultant in the cell comrade Melnikov, experienced in jurisprudence and in many other examples of holding inquest those times announced to me that in the absence of my evidences, the investigative department wanted to make use of persuasions of deputy people's commissar and get false evidences. Therefore, there was no guarantee that the deputy people's commissar would carry out his promise, and, moreover, my conscience would have pricked me for the whole life that I had given false evidences. If the anti-Soviet anecdote invented by me would have concerned only me and nobody else, but the investigator required that I indicated the name of the person whom I had told that anecdote to, i.e. had invented and accused other innocent people except for myself. Shortly speaking, I refused false evidences and didn't write any accusations neither against myself nor against someone else. My companions approved my behavior and my conscience is clear. Andrey N. Tupolev also approved it, when I told him that story in December 1940.

I rejected all accusations, including evidences of the previously arrested (in the end of 1936) B. Dyagterev, who had studied and worked with me at the same department. B. Dyagterev had signed evidence, false all to pieces, that on July 15, 1936 he had had a talk with me in the room of the main engineer and had recruited me for a counter-revolutionary activity. B. Dyagterev rejected these evidences at the Military Board and, notwithstanding this, he was shot by the order of special meeting of P.C.I.A. of the USSR. In 1956 B. Dyagterev was posthumously rehabilitated and restored in the Party ranks. According to all investigation papers I was to be released and P.C.I.A. was to apologize to me for all the abovementioned. But the illegal act happened: on April 23, 1939 I was condemned by a special meeting of P.C.I.A. of the USSR for 8 years in a work camp. I was sentenced in my absence with the formulation "...for active connections with the enemies of the state", meaning previously arrested comrades and specialists whom I had guarded (in the period of my work at the plant and after my arrest).

I am writing this for the further generations in order to avoid repeating that ever again. In order that everyone knew what high Soviet patriotism was required from Andrey N. Tupolev: he not only defended the honor of Soviet specialists, but managed to convince the authorities that all specialists should work for rapid creation of new progressive samples and that time should not be lost for the creation of fabricated, false cases and provocations which had taken years of creative work of Soviet people.

Camp

I would like to quote some fearless references for that time and in that situation about A.Tupolev, given by the well-known Soviet military leader Paul S.Rybalko. He became world-known being the chief marshal of the Tank troops during the Patriotic War (World War Two) and twice the Hero of the Soviet Union. I became closely acquainted with Paul Ribalko in a Pullman freight car, in the beginning of May 1939 at the Moscow-Kazan goods railway station. I and three condemned civilians (aged 28-32 years) were brought to the station and put into the big car of the troop train. We were sent to Krasnoyarsk; then, as it became known later, to Dudinka, in a barge in the Yenisei river and to Norilsk by a narrow-gage railway. 32 soldiers of middle and older age condemned for 15-20 years each were put into that car. Comrade P. Rybalko, condemned for 20 years, was among the military men. He had big authority and influence over the solders and soon gained affection of everybody in the car, as well as in Norilsk camp by his sincerity and warm nature.

Another 4 civilians condemned for 15 years each had been put into the same car before the troop train moved. Comrade Melnichuk was among them; he had been the head of people's commissariat "chief gold" (in his time he had been secretary of the Central Committee of Komsomol, when comrade Chaplin had been the General Secretary of the Central Committee). On the whole, there were 40 people in our freight car.

Our goods troop train speeded only at night, during the daytime it stood at deadlocks at big stations. Thus, it took us about 11 days to get to Krasnoyarsk. During that time we all got acquainted with one another despite of the cram and difficulties of transportation. A bed in the car was a two-storey plank bed of board lumber and hay. The oldest among the military men P. Ribalko was involuntarily elected the head. He could cheer up and reassure anyone; he considered that a great state mistake had taken place on a large scale and that everyone should keep health and courage in order to be able to prove Soviet patriotism working. He assumed that the camp was much better than a jail casemate. P. Rybalko was telling how the investigator had been boasting that they had found that Andrey N. Tupolev had sold drawings to the German aviation company Messerschmidt. When P. Rybalko had called that a scurrilous libel and provocation, all military men and civilians participating in the discussion had agreed with him and had been surprised that the organs had allowed such a provocation. P. Rybalko added that he had been similarly accused and condemned as if for parricide and close contact with "parricides" - marshals Tuhachevsky, Uborevich, Yakir, etc. (rehabilitated posthumously in 1956). In fact there had been no parricide; on the contrary, P. Rybalko had put many efforts in the creation of Soviet tank armies. After the shooting of the innocent comrade Tuhachevsky and other marshals of the Soviet Union because of enemy provocation, all the ideas and suggestions of marshal Tuhachevsky had been declared hostile and unpractical. Those who had supported and insisted on realizing these ideas, as comrade P. Rybalko, had been in danger of disfavor and repression. Only later during the Patriotic war of 1941 the life forced to create aviation and tank Soviet armies.

I met many Soviet citizens - political prisoners, both civilian specialists and military men, all the way long to Krasnoyarsk, and then in the transit point on the bank of the Yenisei river, near Krasnoyarsk and on the barge in the Yenisei river to port Dudinka, as well as at the railway platforms to Norilsk and in the camps of Norilsk. They expressed their concern and anxiety with regard to the defense of our country in connection with mass repressions of Soviet professionals especially in 1937 and 1938.

As it was found later when I got closer acquainted with the political prisoners, they were authoritative specialists (geologists, miners, geodesists, builders, journalists, party and political figures, military professionals mainly from technical branches - artillery, tanks, field-engineering, etc.), some of them were in high ranks - professors, generals (there was no general rank those times, but there were brigade commanders, commanders of corps, etc.). Some of them were close to governmental circles and were informed about the opinion of the Party and Governmental leaders on important questions, pertaining also to the imprisoned specialists. In particular, it was known that comrade Zavenyagin had been appointed the head of the construction of Norilsk industrial complex. He had received special assignment from comrade I. Stalin: industrial complex should have produced the first lot of nickel not later than in 1940. And "...if it happens later, have only yourself to blame". In order to fulfill this assignment comrade Zavenyagin had received significant rights and advantages for the imprisoned: Red Army ration and clothing. Comrade Zavenyagin declared the imprisoned the imprisoned about that in Krasnoyarsk before moving to Dudinka by barges. What was more, comrade Zavenyagin assured that each imprisoned engineer and technician sent to Norilsk would be used in accordance with his profession, in the interests of rapid construction of the industrial complex, that he would be personally meeting each engineer in order to decide how it would be better to apply him by his profession.

Thus, on arrival to Norilsk almost all engineers, including myself, participated in general works for no longer than 2-3 weeks, except military engineers who were allowed not to work because of their diseases. Then the engineers and technicians were brought to the ECW's (Engineer-Construction Worker) barrack with two-storey plank beds, but "cultural conditions of life".

Comrade Zavenyagin had talked with me in detail before I was transferred from general works (I participated in the creation of foundation pit in permafrost soil - we gouged holes, and the shotfirers blasted). I replied many questions: about my education, profession, place of work and position (and also who had been director, chief engineer, and chief instructor at the aviation plant N 22). I told that three directors had been arrested at the plant N 22 one after another, including the director who had been transferred from plant N 39. I also told that after his arrest O. Mitkevich had been appointed director of plant N 22; in four months she had been arrested, too.

Comrade Zavenyagin offered me to work as a designer in a design office/design bureau of the Norilsk industrial complex in the department of non-standard equipment. I was satisfied, because there was no other design work, moreover, in aviation. Besides, comrade Zavenyagin informed me that in March 1939 A. Tupolev had suggested to create a military airplane in jail conditions. This airplane would have to surpass the best airplanes in the world by its qualities. Government of the USSR had ordered to organize aircraft Central Construction Bureau (CCB) and CCBs for other industries using the imprisoned specialists. Comrade Zavenyagin even expressed regret at the fact that with my working experience and good references I hadn't been sent to the CCB of aircraft design, but had been sent to a distant camp, where there had been no possibility to use my profession. I replied that I had been offered to work in that CCB as prisoner under condition that I would sign any false accusation against myself. But since I hadn't signed any accusing compositions neither against myself nor others, as I hadn't wanted to mislead the organs, the party organization, my relatives and close people, for this true relation, the investigators and even deputy people's commissar Kravchenko had declared that I would be lost in the camps with a spade. In reply comrade Zavenyagin shook his head and said that in Norilsk there was also an object, very important for the Fatherland and interesting for many specialists. He said that the efforts of specialists would be correspondingly compensated and thanked, down to the decrease of jail period and early release. Great humanism of comrade Zavenyagin as the director of construction and the head of Norilsk camps created hopes for better future among the prisoners.

We were fed well, mainly with fish products; a bucket with coniferous infusion against scurvy was kept in the barrack; bread was given 800 grams a day per person, mainly brown (rye), some times grey (white bread). Every week or once in 10 days we were brought to the bath-house (banya), clothes were also changed. We wore warm cotton wool trousers, sheepskin short fur coats, valenki (felt boots) and cotton wool caps. But the majority had patched outer clothing (used). A radio loud-speaker was set in the barrack. A senior man and person on duty brought newspapers (a week late) and periodical books, which were given on receipt. Earned money was not given out in cash, but it was allowed to purchase perfume and some food products for a certificate in a stall or little store if the earned or transferred money was present on the account. Many imprisoned, including me, received parcels from relatives. The imprisoned arriving from other camps were satisfied with the rules at Norilsk camps, as elsewhere they were worse.

Geologists found a big veins of nickel, copper, platinum, gold, and other less-common metals. In the ECW's barrack it was decided to suggest comrade Zavenyagin to prepare a lot of nickel and other valuable minerals as a birthday present of I. Stalin - December, 21, 1939, the sixtieth anniversary. The suggestion was accepted and the present was given. We heard that comrade Stalin was very glad that the order had been fulfilled ahead of schedule (in 1939 instead of 1940) and the powerful industrial complex had been created. Obviously, due to all that comrade Zavenyagin was appointed deputy people's commissar of the P.C.I.A. of the USSR on especially important construction objects keeping the position of the head of construction of the industrial complex and the chief of Norilsk camps.

Sunday had always been the day of rest, as well as May and October holidays and the New Year. When it was - 40oC (- 40oF) and snow- storming we were also kept indoors and weren't doing exterior works. Those days we were meeting with friends and exchanged opinions on the further personal and collective fate. Many of us realized that the potential insidious enemy of our Soviet country was the German fascism, headed by Hitler, and his active allies were Italian fascists headed by Mussolini, Japanese militarists and reactionary imperialistic circles of England, France, USA, and other capitalistic countries.

Fascist Germany, having received the green light and direct support from the side of rulers of England, France and other capitalistic countries, occupied the closest states - Austria, Czechoslovakia, and other small countries and established there the fascist regime. Germany started active preparation of attack against our country of Soviets.

The military men, including comrade P. Rybalko, basing on some radio information, made prognoses for possible large military operations and discussed that it was necessary to organize a thorough preparation. These discussions and arguments were interesting because a particular situation was evaluated and questions were put, for example: "Who will be creating the best in the world military bombers and landing airplanes? Who profited that A. Tupolev with a big group of colleagues-constructors were kept far from their direct business for two years, imprisoned, under examination? And do you know how many leading constructors of artillery with great experience were imprisoned and condemned because of slander? Why the tank of the chief constructor comrade Koshkin is considered to be harmful (the future tank T-34)? Why the development of short and long-range rockets was stopped? Where are experienced military leaders, who will be mastering new equipment, who will put questions on its creation before the heads and suggest to the designers of military objects? Finally, who will benefit from the destruction of skilled workers and Soviet patriots? Or their isolation even for a short period of time by slander is to the good of enemies of the Soviet Fatherland." This was a special preparation to the big open war from the side of enemies of our Fatherland. It was necessary to help the leaders of the Party and the Soviet government to restore the skilled workers, specialists, especially those who had gathered experience, and to create young talented personnel. Practically all imprisoned specialists, Soviet citizens condemned by a political clause because of slander and provocations were concerned about that.

I wrote more than 400 petitions for my rehabilitation, but received no replies, neither did the other comrades. Apparently the third department of the Norilsk camp didn't let those petitions pass. Letters from relatives came with lines wiped off, and letters with petitions on rehabilitation reached relatives in the same form.

During winter and spring time all staff of the design bureau was taken to clean the houses and streets from snow. The four-storey houses were under snow up to the roof, we broke passages in the snow, first of all to the doorways, and in the streets.

Hydroplane

A hydro aerodrome was created in Norilsk for the summer period, and as soon as frosts came and snow appeared, the aerodrome took planes with skis. These were low-power, one-and twin-engine planes, which enabled to easily change the skis for floats. There I met Tupolev's construction: twin-engine float (or skis in wintertime) plane of "R6" type. I knew it very well, since it was manufactured at our plant and I had a chance to fly in it as a flight engineer.

This took place in the beginning of summer 1940. I was waken up at night, brought to the person on duty at the camps' department, fed, shaved, and brought by 8 a.m. to the office of the head of the camps. The latter informed that plane "R6" had crashed and that not less than three days were required to find out the causes of the crash, reconstruct the plane and transport products and medicines on it to lake "Lama" (about 120 km from Norilsk), and bring all bed-patients and sitting patients to Norilsk. We heard radio information from the lake that they had no food and that in 3-4 days the badly diseased would be death-threatened. A plane from Krasnoyarsk was promised in about 5-6 days. The only type of transport from Norilsk to lake "Lama" in the summer time was a float plane.

I and other specialists should have given active technical help to fulfill that task in time and comrade Zavenyagin earnestly asked to accomplish this task, promising all administrative, organizational, and personal assistance. Then we were immediately escorted to the hydro aerodrome in a trolley by the narrow-gauge railway. A pilot - a morose young guy, tall, with strong constitution - got into our trolley on the way together with a soldier escorting him. As it was found, the pilot was under house arrest accused of an "attempt of diversion", which led to the crash of plane "R6".

It was found out that there was no passport for the plane "R6", the floats had jagged holes and practically no bottoms, the undercarriage components and joints were in small cracks. An anchor was lying at the plane's float; it was dear to me, designed lightweight by my suggestion. The plane had no armament, but the equipment was transposed for local reasons in order to create a passenger cabin for 5 people. As it was found, the head of the aviation service and pilot had received an urgent order to create a 5-passenger cabin on plane "R6" for the flight to lake "Lama". 3 people from the people's commissariat commission on the perspectives of development of the Norilsk industrial complex, deputy people's commissar Zavenyagin and his daughter had been among the passengers. As there had been no passport for the plane, the head of the aviation service had sent a radio signal to Krasnoyarsk in order to get information on the transposition of the equipment, but had received no reply. So, the equipment had been transposed at his discretion and seats for passengers had been added.

The plane had taken off with passengers at full force speed of engines, it had changed to cruising speed and had gone down. It had not reached the aerodrome and had had to land at shallow places in the Norilka river. The crash had taken place right there, but all the passengers and the crew had been saved.

Airplane "R6" of Andrey N. Tupolev turned out to be safe, sturdy, mainly the floats were damaged. Cracks in the steel longerons of the center section were dangerous, and in order to make sure of that, the corrugated boarding should have been taken off the clinchers. But it was decided to reject the latter variant, as there were no necessary instruments and not enough time. I made everyone sure that the longerons were safe, and we made a trial flight. The plane was quickly restored (a spare float was found, the second one was repaired). It was decided to make a testing flight without steep turns, with careful landing at minimal speeds. But the flight engineer refused, motivating that there might have been cracks under the boarding. It would have taken us long time to open the boarding, and we wouldn't have been able to fulfill the task in three days, which would have led to the death of bed-patients at lake "Lama".

I offered my assistance: to make a trial flight as a flight engineer and mechanic. I explained that I had flown plane "R6" at the plant and knew the material part and technical documentation well (I had been going to be a flight engineer - investigator of heavy machines, but hadn't been allowed to because of bad vision of the right eye), and that I was ready to fly together with the pilot. There were no other interested except for the pilot and myself. The head of aviation service and hydro aerodrome ran to make a call to comrade Zavenyagin, who allowed me to try plane "R6" in the air as a flight engineer together with the pilot, but ordered to put parachutes on all the participants of the flight. During the trial flight I had to get in all the points and pull up the cable control. In about an hour I entered the pilot cabin and showed him a big finger up - everything in order; the pilot, his face shining with joy, showed the big finger, too. In about one hour and a half after taking off the pilot landed the plane smoothly at a minimal speed, classically, as it was acknowledged by everybody. After that the products and medicines were brought without delay to the lake. And after 5 flights 18 patients were brought from there.

We fulfilled the task within two days and 20 hours, i.e. in 68 hours, and we were all especially glad that all patients from the lake had been saved from death. Comrade Zavenyagin thanked all of us for that action and announced that he had solicited for my early release and annulment of conviction.

We were given several days of rest and then were sent to work at better working conditions. Civilian comrades treated me with kindness and offered assistance. Particularly, the pilot, fully rehabilitated on the case of the crash, visited me several times, always in good temper, recalled details of the story of saving people from the lake, telling the chief engineer and other civilians about that.

A month later, approximately in July 1940 I received a telegram from the people's commissar of P.C.I.A. of the USSR that my case had been sent to be reinvestigated for rehabilitation. Approximately in September 1940 the verdict was that my case had to be stopped because of absence of corpus delicti, i.e. I was to be rehabilitated. But, unfortunately, I was brought to Moscow in the end of November 1940, and from October 1940 an order of the people's commissar of P.C.I.A. of the USSR on delaying reinvestigation of cases until special instructions was issued.

Comrade P.Rybalko, who was at our camp department, said that for saving people one should be immediately released and awarded with the order of rescuers. At that time the camp was filled up with the imprisoned brought by sea (from Murmansk city and other places). There were pleasant announcements that all imprisoned specialists would be used in accordance with their profession, that Andrey N. Tupolev and Vladimir M. Petlyakov successfully worked together with other imprisoned constructors - airplane specialists at a special object in Moscow, and that former imprisoned professionals of different fields (motorists, tankers, artillery-men, miners, dressers and other fields) had been already sent to professional work. Particularly, designers of missiles and jet engines comrades V. Glushko, S.Korolev and others were repressed and condemned, but weren't sent to camps, they would work at special objects. Positive decision on the application of military specialists in their profession as consultants and direct executors with previous reinvestigation of cases was being awaited.

To Moscow!

Finally, in late September 1940 I was brought for the reinvestigation of cases to Moscow with a group of 11 comrades, three of them military men (generals), including P.Rybalko. In the port Dudinka we were put on a barge going to Krasnoyarsk city with the imprisoned patients. The Yenisei river started to cover with studge ice, making the way up the stream very difficult for the barge; a towboat was used, so the way was long and we arrived to Krasnoyarsk on day 33rd.

We were given place in the bilge, in the corner lower plank beds. All of us, 11 people, were friends and could stand up for one another against brigands and thieves-recidivists, who penetrated among the patients brought to the continent (before the medical commission recidivists had smoked tea and had drunk chifir (tea brewed with very little water), badly affecting one's heart, which had been the reason of "writing them off" as patients with bad heart).

In two days after the beginning of our voyage on the Yenisei river, late at night four criminals escaped from the deck works. Two of them were caught and the other two sank. After that case confidence in criminals was lost, and we, 11 people, were brought for the deck works (after the agreement with the corresponding authorities by radio). We were woken up very early in the morning and brought back to the bilge late at night. However, it was our pleasure to stay in the fresh air and enjoy the rich and beautiful nature: taiga with picturesque high pines on the right steep bank of the Yenisei and field with a gentle slope on the left bank; sometimes small islands met on our way and very seldom - sparsely populated villages. We were provided with warm clothes: cotton-wool trousers and coats, valenki, short fur coats and cotton-wool caps with ear-flaps, which was the result of personal care of the head of Norilsk construction and camps comrade Zavenyagin.

An imprisoned Soviet citizen stayed in separate deck-cabin. He had been a secret service man and had successfully fulfilled tasks in the favor of the USSR being on the territory of England, Holland, France, Spain, and then Germany, where he had been the organizer of an important Soviet secret service. But in the end of 1937 he had been called to Moscow, where he had been arrested and condemned for 20 years (on the basis of false provocative materials of Hitler's secret service). A woman-doctor stayed in the second cabin - the former wife of the condemned marshal of the Soviet Union comrade Yakir, shot together with marshals Tuhachevsky, Uborevich, and others. She was condemned to 10 years in work camps for her husband, but didn't speak badly about him and earnestly asked the soldiers not to question about her husband, as she wouldn't be telling anything, which conformed to mutual interests. The soldiers together with comrade P. Rybalko were carefully trying to find out several characteristic details, and only in about a month before the arrival to Krasnoyarsk she claimed that she couldn't say anything bad about her husband, discrediting both the Soviet citizen and Soviet military marshal, and that she knew only good things about him. Being a doctor she helped the imprisoned patients and was mainly silent.

The international secret service man was a lively, talkative person, he told many episodes from his life in a clever form, and always warned that to keep our mouths shut. He was a man of great erudition in many spheres: mathematics, physics, Russian and foreign literature, music, painting, sculpture, history, he knew details of significant historical events and personal life of the well-known figures. What was more, he knew seven foreign languages, 5 of which as perfect as Russian. It was interesting to talk with him and it was a pleasure to listen to him, as he literally represented a live encyclopedia, giving even dialectic analysis with Marxist-Lenin substantiation. He was interested in each of us, as well: biographical data, causes of repression, prosecution of investigation. He carefully compared different periods of work of the Soviet organs of State security and periods of activation of subversive activity of foreign intelligence services, mainly English, most qualified, then German fascist Hitler services, which had outdone the English in subtle methods of undermining and provocative activity. At the same time the preventive highly qualified activity of our secret service, which had previously been considered the best in the world, started to weaken. There was no full trust in our secret agents from the side of our government, which led to partial loss of a highly qualified experienced Soviet secret man, beginning from the middle of 1935. Already then at that time our intelligence service knew about the plans of Hitler's secret service, organized in close contact with the English, to eliminate big specialists of different spheres of the Soviet state by all means. In 1936 and 1937 this plan was developed by way of throwing provocative denunciations and false information to the Soviet home front from the side of Hitler and English secret services. Soviet secret agents, particularly he, who worked in Germany, France, and Spain, learned that provocative reports had been prepared. Those reports stated that some of our best specialists, originating from intelligent noble families, first of all, non-party men, had been allegedly recruited by foreign capitalistic secret services and had become "parricides". Several other variants of provocations were added to that, particularly, against Andrey N. Tupolev that he had allegedly sold drawings of fighter-interceptor to Messerschmidt company. It was stated that many other big and middle Soviet specialists had supposedly been trying to become defectors during their business trips abroad and had passed secret data on the Soviet defense to foreign companies. It was also mentioned that the whole group of the well-known specialists were sabotaging and spying and that missile specialists were doing an unnecessary and sabotage work, as in the highly developed countries the missiles allegedly inferior to artillery shells hadn't been produced. That time absolutely secret works on the creation of long-range missiles were started in Germany headed by Von Braun. Our Soviet intelligence service was actively working in Germany, and it had particularly become known that Hitler's secret service had had a full "dossier" with photos of many of our Soviet intelligence agents, who worked in several capitalistic countries. 42 residences of Hitler secret service were implanted in the Soviet Union. Our secret agent received coordinates of their location, with the characteristics of Hitler secret agents and messengers, located in the USSR. He had time and again sent all those data together with the provocative plans of foreign secret services (by secret encryption) to Moscow to particular authorities. But, apparently, our intelligence agents hadn't been much trusted. Three Hitler officers followed him, changing clothes, and our secret agent had made sure that there was a "dossier" with a photo witnessing against him as well.

We treated one another with particular kindness, and each of us was thinking and asked advice of how to help the Soviet Fatherland in strengthening its power and, first of all, defense. The soldiers, as well as the secret agent, were telling many episodes from their life and the life of important comrades of Lenin in the period before and after the October Revolution, which strengthened our belief in ourselves.

In 32 days our barge was brought by towboats to Krasnoyarsk. This was in the end of October 1940. We were sent to the so-called "stage-transit" jail, where the soldiers were separated from us, and the secret agent was brought away, too. Thus, I was sent to Moscow with 7 civilians, all were engineers in different fields of defense (artillerists, chemists, etc.).

Engineer Camp

I and 7 engineers, condemned of a political cause (as though for sabotage, which was denied by the condemned engineers) were brought to Moscow and placed in the Butirskaya jail, and early in the morning after sanitization were put in the cell. I was told that those who had been in that cell had been sent to work in accordance with their profession. By 9 a.m. I had been brought to the deputy people's commissar of P.C.I.A. of the USSR Kravchenko. At the reception I was kindly told that in 15-20 min the deputy people's commissar would invite me.

During our talk the deputy people's commissar asked me if I was willing to work in my profession as aircraft constructor, and I replied that I certainly willed. I told that according to the telegram of people's commissar I was sent for retrial and rehabilitation of deputy people's commissar and the head of Norilsk industrial complex comrade A.Zavenyagin and showed the telegram.

Deputy people's commissar Kravchenko replied that all retrials were postponed until autumn 1940. "It's your right to hand in an application on rehabilitation, but if you are professionally working at the special object as active as you participated in the reconstruction of the plane after the crash, it is safe to say that you will rather be released for the fulfillment of state order at the special object than your case will be reviewed. If I'm not mistaken, I was talking to you in March 1939 in the period of investigation, now the situation is different; I wish you successful work at the special object." I confirmed my wish to actively work in my profession in the sphere of aircraft construction again, after which I was brought to the Butirskaya jail. In about 2 hours I was called out with my personal belongings. An openly glazed bus brought me with two attendants in civilian uniform along the main streets and squares of Moscow (according to my request). I was looking out of the bus windows with enthusiasm and got a great pleasure of seeing my native city in the wonderful frosty weather. On the way the attendant asked my sizes of suit, footwear, and hat. I was brought to the special object on Saturday after the end of work, and when I was brought into the hall on the ground floor I heard familiar voices from the big group of comrades standing in line questioning: "Korotkov arrived: where from?". Those were comrades from our plant whom I had known well for several years since 1927. While I was answering questions, Andrey N. Tupolev approached me with rolled linen under his arm, greeted me by family name and asked to recall family names of condemned comrades - aviation specialists, whom I had met in the camp. I told the names of the comrades whom I had heard about. A. Tupolev thanked me for the patriotic activity in Norilsk. I was especially glad that the location of comrade I. Kostkin and other comrades became known and that all comrades - aviation specialists would be called for the work at the special object and aviation plant. Andrey Tupolev announced that he had requested me to work with him and asked me to apply for that to the head of the special object.

I paid attention to the fact that all comrades were holding linen rolls under their arms. It turned out that Saturday night they were waiting in line for the shower room, where there were 5 showers for maximum 10 people. But that wasn't confusing for anyone, as everything was scheduled for hours and minutes. The same day the administration of the special object provided me with clothes: a model suit, shoes, socks, a shirt with tie, a hat, and light overcoat.

Even in those difficult conditions of imprisonment Andrey N. Tupolev paid great attention to gathering and growing highly qualified specialists -constructors, technologists, and production workers in order to create the most perfect airplanes. In particular, A. Tupolev had convinced corresponding authorities that all imprisoned young specialists should have been used in their profession, the Soviet Fatherland and many specialists were grateful to him for that. All specialists working at the special object called Andrey N. Tupolev father, many old men called him the father of the Soviet aviation, heavy aircraft construction, and of the best airplanes in the world with the highest carrying capacity and speed.

A group of 25 comrades headed by V. Petlyakov, the first student and companion of Tupolev, was released early due to the creation of "P-2" high-speed diving bomber at a special object in 1940. They were released from conviction and each was awarded with 25 thousand roubles, and comrade V. Petlyakov - 50 thousand roubles.

I was sent to work on the creation of a high-altitude bomber with three pressurized cabins under the direction of Chief designer comrade V. Myasizhev. A. Tupolev told me that despite of his requests I was sent to work with comrade V. Myasizhev, since the new machine only started to be developed and there were not enough personnel, and assembling of airplane "101" ("TU-2") of A.Tupolev had already been started. The head of the special object Kutepov announced that specialists who participated in the development of working drawings, production and testing of the airplane would be released early for the creation of the new airplane for serial production.

Andrey Tupolev demonstrated special sensitivity and support to me, particularly he showed that he didn't have full power and authority. Therefore, it was important to be patient and strong enough (we all possessed that) and work on the creation of "102" machine of V. Myasizhev, our country also needed those machines. Andrey Tupolev promised to keep me in mind in the future.

Notwithstanding a great amount of his responsibilities, Tupolev loved and actively supported new ideas and fantasies, often talked with us in the evening hours and was an example for young people literally in everything, starting from morning gymnastics, body rub-down (applying cologne when the shower with hot water was missing), and evening walks on the roof (where there was an additional brick floor for the library and chess and the walking area covered with metal gauze from the sides and from above). I, Leonid Dyakonov and Anatoly Bardin had to do exercises on the walking area with Andrey Tupolev several times, and meeting him after his release he was joking "Haven't you forgot the physical exercises?", and after our positive reply he said: "Very well, the work goes better after physical exercises, for spite all the enemies".

S.P. Korolev

Andrey Tupolev specially supported the ideas with a perspective of their realization. One of the bright examples was the case in December 1940. Andrey Tupolev recommended comrade Sergei Korolev to work on the creation of long-range missiles with reactive engine which would play an important role in the future. The case was that S. Korolev, being convicted because of enemy slander as a saboteur of aircraft construction, time and again asked the heads of P.C.I.A. of the USSR, writing letters to comrade I. Stalin, to be allowed to develop the reactive long-range missile. He wasn't allowed to, motivating that the topic wasn't interesting, as nobody was doing that abroad in progressive countries, that this fantasy wasn't necessary. Then S. Korolev, by advice and with support of Tupolev, started doing calculations and drawings of different variants of long-range missiles in the evenings until late night and sometimes until Sunday morning, and during working hours carried out the orders of the chief constructor comrade V. Myasizhev on the creation of pressurized cabins of plane "102".

S. Korolev spent much energy for such hard work and it was difficult to take him for evening walks. When needed, we helped him to build calculation graphs. Some comrades advised Sergei not to work in the evenings and at night wasting time on "fantasies", but devote only weekend day hours to it, the others recommended to give up the "missile fantasies", as the heads of the special object were not interested in that topic and could complain about wasting forces at night for irrelevant activity. But S. Korolev, feeling great support from the side of A. Tupolev continued doing research in his personal time, using theoretical basics of Ziolkovsky, his own theoretical calculations and tests of prototype models, which he had carried out before the arrest, on the creation of powerful reactive engines and long-range missiles with unbelievably high speed for those times.

One early Sunday morning Sergei Korolev entered our bedroom it was March 1941, his eyes were sparkling, and he informed us that the speed 5000 km/h could be solved by the capabilities of techniques and that it was necessary to work further. (In the nearest future Sergei Korolev would design the first Russian rocket system - "Katusha" - which was successfully explored against fascists during the war. After the war he would be the General Designer of Russian space-ships - from the first "Sputniks" to "Vostok" operating even in the XXI century. By 1941 Korolev was imprisoned and forbidden to work on the rocket design.)

Andrey Tupolev paid attention to his tired face with blue streaks after intensive work at night and he rebuked Korolev for excessive and tiresome work at night. Tupolev also stressed that the topic of long-range missiles was very prospective and that its importance wouldn't be quickly understood, therefore it was necessary to work on it, but moderately, not damaging health. (Sergei Korolev was 34 then).

The beginning of the Patriotic War of 1941-45 broke all plans and it was necessary to rapidly evacuate military and other plants to Ural and Siberia areas and without delay put to rights military serial production. On the third day of the war, Andrey Tupolev had an early release and was freed from charges. Comrade I. Stalin personally expressed his thanks and presented a money award to Tupolev for the creation of the world-best diving twin-engine bomber with large bomb-load, high speed and large-caliper machine-guns.

We were all rapidly transported with the equipment to Omsk in Siberia. In Omsk a group of more than 30 imprisoned specialists had an early release, as well. Their families were brought to Omsk at their request and were sent for work, accommodation was provided.

Our Soviet army was fighting and retreated under the superior forces of Hitler troops, many cities with aviation plants remained in the enemy's hands, and the plants of the cities' central districts suffered intensive bombardment. There were not enough fighters and bombers with modern equipment, since planes of 1930's were inferior in military equipment and despite heroic fights of our pilots, many of them perished. New modern airplanes in mass quantity were urgently required.

There were no aviation plants in Omsk before the Patriotic War. Only in the end of June 1941 air-engine Zaporozhsky plant was evacuated and it was necessary to build a large aircraft plant from scratch with subsidiaries for the creation of equipment and arming, and, moreover, to build a pilot plant for the creation of pilot airplanes, which had been planned in Moscow before the Patriotic War.

The process of construction of the plant began with active participation of the imprisoned specialists, colossal trust was shown to us -it was successfully helpful, and from November 1941 we started producing the planes TU-2 (diving) and fighters YAK.

Sergei Korolev actively worked on the development of serial drawings of the airplane TU-2 and organization of shop manufacturing, as well as gave technical help on the production and testing of pressurized cabins of plane "102" at the open plant. At night he continued working on the drawings and calculations of reactive missiles, striving for being allowed to work on the creation of such redoubtable weapon. But his request was rejected and he was forbidden to work at nights, as that affected the jail order, but that prohibition did not influence Korolev. Then he was warned that he would be sent to the work camp for breaking the jail order (it was required that the imprisoned stayed in his bed and slept or ate for 23 hours). After that we should have taken immediate measures to save Sergei from camp, as before that the head of propeller crew of plane "102" had been sent to the camp and the chief constructor comrade V. Myasizhev, whatever measures had been taken, couldn't have saved his comrade from deportation to camp, and he hadn't been found in the war period.

We found a variant of saving Korolev: to transfer him to the position of deputy chief of the fuselage shop comrade Heller (imprisoned), who was the initiator of that proposal. Comrade Heller was a specialist in different kinds of stamping and a good administrator (he had previously worked at Gorky aviation plant as the head of stamping department, before that he had graduated from the working faculty of Leningrad Polytechnic Institute, and had been on a business trip to the USA at the car factory "Ford"; he had been repressed and condemned because of slander), but not an airplane specialist, so he needed assistance in military equipment acceptance. Coordination with the construction department would take Korolev not more than 5 hours a day. He could spend the rest 6 of the 11 working hours at the chief's office working on the "fantastic" missiles, and sleep well at night. But in spite of the fact that many comrades were trying to persuade him, Sergei Korolev didn't agree to that work even temporarily, as it was not a creative work. And only after the conversation with Tupolev he agreed. Andrey Tupolev suggested working only in free day hours, in the office of the shop's chief; otherwise he could have been put into a camp for breaking the prison order. Tupolev convinced Korolev, as Korolev reckoned him, and the Great father of Soviet aviation took care of him, and valued the future Great designer of space ships, powerful intercontinental missiles and antimissiles, which provided us peaceful work and tranquility.

Tupolev's assumptions came true - in about three months of working in a shop Korolev received an order from the Soviet Government to create the long-range missile; it was in June 1942. That took place after the German VAU-2 had appeared, being launched for the destruction of London. Owing to the moral support of Tupolev, Korolev already had a theoretical calculation of the reactive long-range missile - more perfect than VAU-2, - with the jet thrust to the very target, providing a precise target hit. Advices and support of Tupolev - useful, kind, and perspective, turned out to be very important to Korolev, and especially to our Soviet Fatherland.

Sergei Korolev was brought to Kazan to the air-engine special object, where together with other imprisoned specialists he was working on the creation of jet engines of comrade V. Glushko (a member of the presidium of the Academy of Sciences of the USSR, twice-hero of the socialist work). Jet engines were planned to be set on heavy bombers as accelerators of airspeed. But the first samples of jet engines (liquid type) weren't successful; they required great fuel consumption and big weight. The design didn't allow setting the engines on the planes. From June 1942 comrades Korolev and Glushko and assistants from the group of the imprisoned specialists began to work on the creation of long-range reactive missile with pulling power up to the very target.

Release

On August 18, 1943 a whole group of comrades from the airplane special object was released after successful state trials. They were working on the creation of the long-range bomber with three pressurized cabins, providing the flight altitude up to 14 km. We were awarded with 25 thousand rubles and vouchers for shoes (American), suit and coat, as well as food, "SP" and liter "B" rations.

In 1945 I was sent to Moscow to the experimental-development bureau of chief designer comrade Tupolev for the control of drawings for launching serial production of new four-engine long-range high-altitude bomber: with three pressurized cabins, perfect construction with the best aircraft instruments and armament, a prototype of the flying fortress of the USA air force, which was kept secret and wasn't passed to our air forces.

In 1946 by the order of deputy minister of aviation industry comrade Dementyev I was transferred from Kazan aviation plant of Gorbunov to the command of personnel administration of Aviation Industry Ministry, from where I was sent to the post graduate school of Moscow aviation institute. Then I went to a correspondence post-graduate school and worked at the plant N 240, in the experimental-development bureau of chief designer of the plant comrade S. Ilyushin in the department subject to comrade E. Zhekunov, who was deputy of the chief designer S. Ilyushin on technology and production.

New repression

In 1947 I and other comrades, former party members, were "illegally repressed". 1937 - 39 were repeating. As it was found later, a well-organized attempt of detriment of our Soviet home front was started by reactionary imperialistic circles headed by American and English secret services. They managed to use the hostile and careerist elements within the official circles of our Soviet country and cause certain damage to our state. The "Case" of Leningrad party organization of comrades Popkov, Kuznezov, Voznesensky, etc.; the failed "case" of Moscow party organization; the "case" of physicians-academicians; the "Jewish case"; a total defamation of the old Bolsheviks and former party members previously repressed - all this pertained to that period.

All our "jail cell" friends (cellmates)who didn't join the party or weren't admitted there again, retired from the secret and off-the-record work with the statement "because of reduction of the staff and some comrades were repressed again without cause after release (annulment of convictions). I also retired from the research bureau of chief constructor comrade S. Ilyushin because of reduction of the staff but at my request I was transferred to the experimental design office of passenger helicopters of comrade Bratuhin. But that enterprise was liquidated in 1948, after which I moved to the experimental-development bureau on the development of flying wings with jet engines.

At that period we all hoped for Tupolev's help and he did everything he could including the talk with I. Stalin, with the request to restore the retired specialists in aviation industry, who had previously been repressed (until 1941) and had been working on the creation of new airplanes during the war, for which they had been released they had an early release and were freed from charges. But, unfortunately, some time after reinstatement of employment (in three weeks) many of those comrades were arrested without any causes. That was a demonstration to the other comrades, including myself that we shouldn't have thought to restore in aviation industry.

I and one of my friends, who turned out to be in the same situation, decided to meet with S. Korolev in order to apply for work. Having arrived to a plant near Moscow early in the morning, we were waiting for Korolev near the entrance. Finally, 15 minutes before the beginning of work we met him. He greeted us very friendly, replied "certainly" to my to my request to work with him and told that he was responsible for us as for himself. In case if we worked with the head of another project, he would give good references to us both as specialists in our sphere and as disciplined Soviet citizens, reliable and trustworthy.

Sergei Korolev told us briefly that he was doing well that he was yet the chief designer of the project. In order to achieve full success Sergei Korolev made certain terms on granting him great organizational rights, so that he could actively develop efficient creative work. It was difficult for Korolev to realize his ideas, high-grade and valuable for the defense: he met various obstacles and formal bureaucratic opposition as he was non-party then (Sergei Korolev joined the party only after rehabilitation in 1956, following Tupolev's example; many comrades who had been non-party followed that example before rehabilitation and repression).

We asked Korolev about the problem of "staff reduction - retirement" in their system. He replied that, unfortunately, 1937-39 repeated. Three people were arrested and five people were retired with the transfer to other system from the plant where Korolev worked, but he didn't know those comrades personally. He cheered us up: "Nothing ventured, nothing gained, so let us find good protectors among the authorities of our country showing our fruitful results. Come and join our staff. I am sure that A. Tupolev will manage to find a fair resolution and rehabilitation of comrades who suffered innocently". We thanked S. Korolev and recovered our spirits, having decided to join Korolev's staff. But the events of the next days overturned everything in our minds: many previously repressed specialists, many former members of the Bolsheviks' party, were arrested (on the basis of falsifications and slander) and retired because of staff reduction at many aviation aircraft plants and in the other spheres of the defense industry at the urgent request of enemy forces, settled in the national security controls. Andrey Tupolev got a disease because of severe disorder, and A. Arhangelsky didn't know what to do with the comrades, including myself, as he promised to find the work for us at the periphery in the other cities where specialists for the production of "TU" airplanes were needed. A. Arhangelsky advised to refrain from the recovery in the aviation industry for a while. He and other comrades also recommended to forbear from joining the staff of S. Korolev until the situation changed, otherwise we would be reduced and would let down not only ourselves, but Sergei Korolev, who had big success in the development of new missiles then (1948), but yet weak influence in the official circles.

As I was told, Andrey Tupolev addressed I. Stalin again about saving and growing the aviation specialists and comrade Stalin gave an order to L. Beria (first deputy of the Chairman of Soviet Ministry of the USSR on national security and inner affairs) about saving the staff. The latter invited Tupolev, calmed him and reassured that it was necessary to take national security into account and that he would consider the appreciation of specialists repressed and retired because of staff reduction from the aviation industry on the subject of using them in state structures. That fact, though insignificantly, but positively influenced on saving lives of the comrades who suffered.

Post-war years

In September 1948 I and other comrades had to enter the institute of transport mechanical engineering "Orgtransmash" as senior designers, where we gave good account of ourselves for a short period of time and I was appointed chief engineer of the brigade.

In March 1951 I was transferred to the Leningrad subsidiary of "Orgtransmash". There I also had to work on the rigging for the construction of building machines, steam and hydraulic turbines and from 1954 - coal-concentrating machines (in the All-Union Research institute of coal concentration, renamed "Gipromashconcentration" institute, where I worked until autumn 1972). It was

Georgy Korotkov with his son Konstantin,
November 8, 1954, Leningrad

explained by the fact that my "tail" (previously repressed and non-restored in the Communist Party of the USSR) didn't let me work in the aviation industry or at another security-guarded work. (I was several times invited by the directors, but the "filtering" organs delayed my application for the work in design bureaus).

But what everyone appreciated in me wherever I worked was Tupolev's working style enabling me to develop high-quality designs ahead of schedule which brought big economic effect to the state.

Thus, for example, I managed to create and implement several machines for the concentration of coking coals of light and middle reversibility for Kuznetsk, Vorkuta and other coal basins. These machines three times surpassed the American machines "Air-Flow" in productivity and in quality of concentration. Then I participated in the creation of machines for the concentration of ferrous and base metal ores, including rare and precious minerals. And I am first of all grateful to Andrey N. Tupolev for everything what I have created. This great Soviet patriot made a significant invaluable contribution not only in the sphere of aviation; he created the Tupolev's school, Tupolev's system of development of the new Soviet machinery and Soviet specialists, devotedly loving their Soviet Fatherland.

In 1956 I was rehabilitated and restored in the party with party service from 1927. Thus, I had worked in aviation industry for 28 years and in the other spheres of industry for 24 years (since 1923 in trade union).

I am a pensioner since 1971, labor veteran of the Ministry of Heavy Industry, but work periodically not less than 8 months a year.

My two sons - Michael (a journalist of literature department of the newspaper "Novosty") and Konstantin (researcher on physics at the chair of Polytechnic Institute in Leningrad) love the great specialist and big Soviet patriot Andrey N. Tupolev, a true Leninist of our native Communist Party. All of us dearly love and respect the Soviet aviation company of Tupolev and are sure that this company will continue glorifying our Soviet Fatherland with its achievements.

Georgy Kirillovich Korotkov,
1908-1978,
Member of Leninist Young Communist League of the Soviet Union from 1922; Member of labor union from 1923; Member of Communistic Party of Soviet Union from 1927.
Leningrad, 1975

POLYTECHNIC INSTITUTE

By the 1970's, of course, the situation had already changed. The time of the unrestrained repressions had passed, although the KGB kept handling the country without gloves. It is worth mentioning that in Soviet times the expression "to handle without gloves" (in Russian - "to keep in hedgehog gloves") was associated with Ezhov (hedgehog is "ezh" in Russian) - one of the first people's commissars of the P.C.I.A., who had participated in mass repressions, but had later been destroyed by the machine which he had created. The Soviet Union was a huge concentration camp with strict rules and immediate punishment, but the majority of citizens had no idea about that and enjoyed life, laughing at the stupidity of the political system. The main distraction was anecdotes although it wasn't safe to tell them. Siberian camps weren't liquidated by anyone, they were only re-equipped. That was a well-known part of life. Life under control of KGB, under conditions of total shadowing, empty store shelves and absence of the simplest goods and foodstuffs. When bananas were brought into the city, there were great lines to buy them and they were in the first place given to children. Being in the USA in 1989 I was surprised why people ate so few bananas. However, I love them even now. Maybe this has remained from childhood.

But we lived, we created, we were happy, we fell in love, and many people recall those years as the time of prosperity and material well-being. Not conditions of our life are important, but our attitude to them. Our world is inside us, only the environment is outside.

A wonderful example of the Soviet life style might be the department of physical electronics of Leningrad Polytechnic Institute, where I was happy to work.

The style of every scientific institution was fully determined by the character and scientific importance of its head. In this regard the department of physical electronics of Leningrad Polytechnic Institute was lucky. It was founded by an outstanding Russian physicist Peter I. Lukirsky. Leonty N. Dobrezov was running this department for many years, and in the 1970's it was headed by Professor Nicholas N. Petrov. All these people were good scientists, had serious scientific works and published a lot of monographs. And what was the most important - these were very intelligent and deeply honest people. Under their influence and owing to their personal examples in many respects, a very pleasant and warm atmosphere of friendly and respectful relation to one another, from a venerable professor to a shy student - newcomer was created at the department. We used to work a lot and to have a good rest all together. Never a week passed but a cause for a small holiday at the laboratory appeared: someone's birthday, defense of thesis or completion of work on the topic - everything was a cause for a feast. Now, after many years we recall even such a small event as a visit to a vegetable storehouse with great pleasure.

In the Soviet years the free working force of the urban population was widely used for the maintenance of kolkhoz (collective farm) agriculture. Each plant, factory, firm, or institute had to follow a certain schedule of agricultural works. In autumn - harvest; in winter - processing of decaying vegetables in antediluvian vegetable stores. And each day a group of colleagues in quilted jackets and boots went to do public works. These were hundreds of thousands of people every day all over the country. Only the diseased and very old citizens were dismissed. Party committees were responsible for the organization of those works: starting from the city committees to party committees at firms, departments, chairs, laboratories, and groups. This was an integrated network stretched all over the country, involving all population capable of working and run by a firm Moscow hand. In principle, one could decline compulsory works, but after a few times such a person lost all possibilities to develop a career, receive get bonuses or privilege tickets to health centers. Such people were dismissed at the first opportunity with the corresponding record in their workbook, which practically blocked the possibility to further work in their profession. The only way remained - to stokers or dissidents... Therefore, the overwhelming majority of common Soviet people preferred not to grumble and calmly went to the vegetable storehouses. In fact, there were sometimes "silent revolts." For example, once in a store the customers found accurate cards in packed bags with potatoes. The text on the cards said: "These potatoes were packed for you by the doctor of technical sciences, professor, The State Prize Laureate, Ivanov I.I. Bon appetit!" People laughed, professor Ivanov was interrogated at the Party bureau and... the system continued working well. So, at our department each period spent at a vegetable storehouse or the kolkhoz [collective farm] became a cause for a wonderful feast. In half an hour before the lunch break special couriers were sent to a store to get port and vodka, boxes were put together, were covered with newspapers and became a table, bottles and snacks brought from home were put on this table, after three glasses the border between professor and senior research officer became very thin. Various problems were discussed between toasts: from the situation at Arab-Israeli front to the peculiarities of exo-electron emission of monocrystalline tungsten. Having had a good snack and having drunk a little bit it wasn't that difficult to come back to the dirty potatoes and hold out until the end of the working day.

Joint work and joint feasts disposed to informal communication even behind the walls of the vegetable storehouse. Any colleague irrespective of his or her job or position could come to any teacher or even to the head of department and discuss both scientific topics and his/her personal problems. Senior lecturer N. Ban'kovsky was the most popular in that.

A big man, taller than medium height, distinguished by special portliness and reliability, he was able to comport himself wonderfully in any situation. During the years of joint work I saw Dr. Ban'kovsky loosing his temper and showing anger just twice, he mainly kept his Olympic calm and balance. He had years of war behind himself. He passed it starting from a private soldier, having left home since school-days, to combatant officer. Being commander of a company of signalers, he was injured several times, participated in the assault of Berlin, and finished the battle path in 1945 in Manchzhuria. Dr. Norbert Ban'kovsky had rarely told about his war experiences, had reluctantly visited veteran meetings, but when he once came to the institute on the Victory Holiday in all his battle orders - they covered the coat by a thick armor. The main occupations of Ban'kovsky were teaching and social work. Those years no meeting passed and no decision was taken without party bosses taking part. Life of any institution of the Union was mainly determined by the party committees. Any serious document had the signature of the party bureau secretary after the signature of the head: director, rector or head of department. And any social questions - from a reference to a tourist trip to Bulgaria and to settling disputes between the colleagues - were fully given to the party bureaus and party committees. Such committees appreciated the participation of Ban'kovsky. He brought calm to the excited ranks of the enemies by his steadiness and deliberation and, having studied the circumstances of the case well, found some decision. If he had wished he could have made a party career: a War veteran, candidate of science, drinking little, but he politely rejected such offers and preferred the life of a common teacher to profitable, but anxious party activity. It is well enough that the salary of a senior lecturer, together with payments for the scientific work were quite sufficient for reasonable living at that time.

Ban'kovsky was doing serious scientific work, although that work was only one of the directions of his many-sided activity. There was a wonderful tradition at the department: students were assigned to the laboratories from the third year and spent several days a week mastering subtleties of experimental scientific activity. In five and a half years the education was completed by defending the thesis done at the same place, in the same laboratory. After that the most talented students were kept at the department, in order to continue scientific work. Thus the lads comprehended what was Science in practice, in live communication and apart from abstract lecture knowledge they received real practical skills.

When I first got into the laboratory of Dr. Norbert Ban'kovsky, the best I could do was to repair a broken electrical socket, after some careful thought. I met a wonderful man in the laboratory, Valentin K. Shigalev, all friends called him "VKS". To me he was an example of how one should do any work. When some task got into his hands: construction of a specialized manometer for the measurement of low pressures of helium, development of a new electronic scheme of measurements or simple repairs of an iron for a colleague, VKS sat, retired into himself and investigated the essence of the problem. Having got the problem and its main rubs, VKS found his own novel decision. That concerned both the iron and the scientific task. Therefore, it seemed that there was nothing that VKS couldn't have solved. Looking at him, I also gradually lost the fear of new problems, although, of course, I didn't become such a virtuoso. That was obviously given by God! VKS was the right hand and the first assistant of Dr. Ban'kovsky. Dr. Ban'kovsky found Customers, discussed the essence and peculiarities of working with them, organized financing, and VKS built the experimental setup and obtained results. Writing reports - that was what VKS didn't like at all, so that part of the work mainly fell on the shoulders of Dr. Ban'kovsky. He was a great master in writing papers. I remember that once the day before demonstrating the report the experimental instrument necessary for getting final results burned out, and the work remained incomplete. If that was straightly written - good-bye bonus! Dr. Ban'kovsky went into his office for a couple of hours, and when he came out he read us a sample of casuistical prose: four pages of text with the discussion of work done, from which it wasn't clear if the work was completed or not, but it was clearly brought out that in order to solve the task successfully the work should be continued for another year and a half... Thus everything was moving little by little. Small scientific problems were solved, once in half a year all the laboratory was doing an emergency job of preparing a report or defending another student, articles in scientific journals were published; Dr. Ban'kovsky participated in one or another commission, etc. Scientific work was connected with the development of new cathodes for electronic devices, and in the Soviet Union all such topics were considered to be secret, therefore we couldn't even think about communicating with foreign scientists or participating in the "open" conferences. The West was separated from us by a thick curtain, and the fact that there was no access to secret documents and that just a few subscriptions were allowed cut away any opportunity of live communication with the Western colleagues. Thus, experiments on the Kirlian effect became one of our official subjects. We started from vacuum EPC: our department was the department of physical electronics and all questions connected with vacuum equipment were close and understandable to us. And that seemed to be one of the most intriguing inventions of Semyon Kirlian. I made drafts on the basis of Kirlian's drawings; in a week glass-blowers produced a device; and here came an exciting moment of the first experiment! Having pumped out the air from the device, we started increasing the voltage gradually. First a weak bluish glow appeared in over the entire space of the flask, then at some moment the luminescent screen of the tube flashed brightly and letter "K" lighted up on it - an increased glowing image of a metal letter welded to the electrode. All the "population" of the laboratory gathered to watch the glow. We were looking at the screen perhaps with the same feeling as the inventors of the first television set were watching their brain-child in operation. But it was then found that in contrast to the TV that device didn't have any perspectives: the amplification didn't exceed several hundreds, resolving power wasn't more than a hundred microns, and it was principally unusable for the investigation of biological objects - try to place a cockroach or a leaf into vacuum... But all that was found later. At the start the investigation of this device brought us pleasure. Using the laws of electron optics we managed to write an equation of movement of electrons in such a tube and explain the process of image initiation. The calculation enabled to predict a whole series of new properties and then check them experimentally. In a couple of months of work the first scientific publication was ready . At the same time we experimented with photographing fingers. I and VKS assembled the first setup, locked in a dark photo-developing room, put a metal subject on the electrode and pressed the button. A buzz resounded and a bluish glow appeared around the subject. That was excitingly interesting and attractive. At the next stage we had to put the finger on the electrode. It was quite frightful. 20 thousand Volts! After some hesitation I put my finger and said: "Press". Again the buzz resounded, I felt a perceptible electric shock and fluorescence appeared around my finger. It was unpleasant, but bearable. This was our first contact with the real Kirlian effect.

Then long months were spent in order to find out the physics of this process. To determine conditions under which the finger "didn't feel the current" and yet useful information appeared, to understand how that should be done.

These experiments gave purely aesthetic pleasure apart from the scientific. The glow pictures were wonderfully beautiful, especially studied with the help of increasing resolution. Perhaps the Kirlians were the ones who described that most beautifully in their small book 2:

"Various spark discharge channels perform their difficult work". The channels-giants wildly blaze with lilac-fiery color. Orange and light-blue "dwarf stars" shine calmly next to them, within the "lumps" of the skin coverlet. Why the giants are lilac, and the stars orange and light-blue? And why are they different in size?

Sheet lightnings blaze, too. These are twinkling craters, and not the fiery lava stream from them, but a glow similar to the northern lights. Inseparable yellow and light-blue twins blaze up here or there. Faded medusa-like figures emerge, as if from an underground cave. They wave and swim in the space, looking for the ones similar to themselves, and having met them, flow together with them or hide in another underground cave. Some discharge channels, as if lighting their way with the tongue of flame from time to time, hurry along the skin gorges in a single file. Where from and where to does this caravan go?

Here they are the mysterious toilers of high-frequency field, guardians of enigmas of the live organism, fathers of the world!"

The work was in full swing. Thus, I got two parallel topics for research. Each required time and effort. In order to obtain results in science one needs to maximally concentrate and fully devote himself to the research. At some moment it becomes clear that it is necessary to make the choice. But when you rush forward, full of plans and purposes, it is impossible to stop and decide to completely change the pattern of life. And here comes the point of bifurcation, and if we can't decide ourselves, the decision is given to us by Higher forces.

In that way after some time I found myself in the situation of Buridan's ass. On the one hand, I was working on the problem which interested and fascinated me greatly, on the other hand, I continued my work on the thesis, which was half done and several scientific papers had already been published. Many have the first-hand knowledge: you do some work, spend a lot of time and efforts on it, it already makes you feel pretty sick about itself, you start doubting if it is necessary, but you can't stop, it's a pity to drop it, the process goes on its own; to jump out of it on the move is the same as to jump from a moving train. The process turns out to be stronger than the person involved in it. Thus it was with me, and in parallel with the Kirlian effect I was patiently toiling with the two-component adsorption on polycrystalline wolfram.

BIFURCATION POINT

The science synergetics appeared in the middle of the Twentieth century, the science about open, dissipative systems. In the Nineteenth century it was assumed that any system could be considered as a closed system, living its own inner life, restricted by the borders, covers, and skin coverlet. And all the processes taking place within this system depended on it. All physics has been based on these principles. Thermodynamics was based on this, too. Biology started to be developed on these principles. However, a whole series of paradoxes appeared. And the main paradox was the heat death of the Universe. The notion of entropy was introduced as a measure of disorder in the processes of different nature. And it was demonstrated that entropy always grew in any physical phenomena, i.e. a part of energy received by the system was spent on heating, because the increase of entropy was taken equivalent to the increase of temperature. Hence, particularly, it followed that an engine with 100% coefficient of efficiency was impossible. The calculations demonstrated that our Universe and, particularly, our Earth should overheat and die of thermal death in quite a short period of time. But, as we see, that hasn't happened within billions of years.

The solution of this paradox took half of the Twentieth century. The paradox was solved by the physicists who created a new science - synergetics, the science of open systems. Systems which exchange energy, and, as we know today, information with the environment. The thought about that seems trivial, but the science of the Nineteenth century investigated only closed or loop systems. Such a notion was introduced for the convenience of the mathematical study, and then everybody somehow forgot that that was a mathematical condition, and led to distributing conclusions of the outer world phenomena. In the majority of cases the model of closed systems gave plausible conclusions, being in good agreement with the experimental data. A few paradoxes not explained by the existing science of the Nineteenth century were simply ignored. But they gave birth to the science of the Twentieth century. Particularly, to the science of synergetics.

Any biological phenomenon, including a human being, is an open system. The new science started developing rapidly. One of its founders, Ilya Prigogine, was awarded a Nobel Prize. This science opened a lot of very interesting lines of research not only in thermodynamics and physics, but also in biology. One of the consequences of synergetic ideas is that any process, whether in engineering or biology, passes the so-called bifurcation points or branch points. The process develops according to certain rules, and then it comes to some point where several outcomes are possible, or several ways of development. The process passes to a new phase depending on conditions, maybe even very random conditions. Say, a crisis takes place in the development of a severe disease - a bifurcation point - and the way out might be a recovery, a chronic disease, or death. Graduation from the university is a bifurcation point for a student - then life can develop according to several scenarios: one or another job, a new circle of friends, a new unexpected stage of life.

If we look around, we will see that all our life consists of transitions from one bifurcation point to another one. From time to time we have to make decisions: what to do, where to move, how to develop further. These decisions might come out from long meditation, sometimes they are random, but the result of these decisions is mainly unpredictable. We can only assume, make plans, and life turns everything in its own way: we never know what is waiting for us ahead and what consequences will grow from our actions. But very often all further life depends on these actions.

* * *

Summer 1976 augured many interesting events. By that time we had had a wonderful rock-climbing team. One of the strongest, if not the strongest alpinist team of the Soviet Union. That team had formed during several years under the guidance of a strong alpinist, extraordinary person, professor and doctor of sciences Anatoly Shevchenko. He was a person able to attract people, to inspire them with ideas/to attract them with his ideas, to arouse enthusiasm, and establish order with his iron hand if a situation seemed to be out of control. Each business is moved first of all by the leader. And Anatoly was this powerful leader in the team. He managed to gather not only strong alpinists, but also good people. Each climb accomplished together was not only a technical peak of perfection, but a wonderful time of communication and intellectual entertainment. That was the rarest situation, possible perhaps only when such an extreme kind of sport was at the height of public interest. Each season reporters and officials met the team coming back from the mountains covered with the rays of glory. Solemn meetings and discussions were organized; all that was broadcasted on TV and discussed in press.

I managed to spend two seasons - four summer months on this team. Perhaps, those were the best seasons in my life, considering both the professional level of mountaineering and the pleasure derived from that.

The year 1977 foreboded a similar season. The plans were grandiose. We planned to go to the very heart of the Pamirs and climb the wall which had never been conquered by anyone. Our climb should have become one of the milestones of development of technical alpinism in the Soviet Union.

A couple of months before the departure Anatoly advised me and another participant of the team,

"This time, guys, go to the training camp first. Spend a week with our young alpinists. It's necessary to grow the new generation".

"But why? We need to get ready for the new climb!"

"That's all right, you'll be just in time. We'll come, prepare the base camp, and plan the route. And then you will arrive. A helicopter will take you up."

This way we found ourselves training young sportsmen in the mountains of the Pamirs-Alays. We started a usual alpinist life with climbs and training, breathing in the heady mountain air, often driving to romantic walks with young girl-mountaineers, ending with passionate embraces under the starry sky. In short, this was a wonderful preparation for the sport season.

One day a helicopter landed on the meadow where our camp was based. This could have been some officials or some people from the directorate. When we saw the heads of the Alpinism Federation coming out of the machine, we realized that something had happened - they looked too gloomy.

"What's happened?"

"Take courage, guys. The Soviet alpinism has suffered bereavement".

"What's up?"

"All the team of Shevchenko was lost".

These words were so shocking that at first we couldn't realize the essence. It was like a horsewhip blow, when you don't feel pain from the beginning.

"What? How could that have happened?!"

"The car with the whole team fell off a precipice. Only two people survived."

As we later learned, an unskilled driver put the car in neutral on a slope, which didn't seem to be steep at all, but was quite long. The car started to gather speed. When he tried to throw in the clutch, the gear-box was practically pulled out, as the speed was very high. The driver still had a chance to jump out on the slope and reduce the speed. When he tried to do so, a tank truck appeared from the turn. The driver twisted the wheel, and the car crashed down into the precipice at the speed of 140 km/h, somersaulting down a steep slope and throwing out the guys sleeping in the car. Most of them died on the spot; only two of those brought to the hospital were rescued.

Until now, after many years, I keep asking myself a question why it was me whom Anatoly decided to send to the training camp?

Life gathers many points of bifurcation like that. We take a decision, and we can't predict what comes out of it. One of such points was connected with my work on the Kirlian effect.

HAND OF DIVINE INTENT

On February 2, 1978 we went to a summer cottage with friends to celebrate someone's birthday. We drank and ate a snack, then drank more, had another snack, then drank again, went for a walk in the night winter forest, drank more during the walk, and when we came back to the house, one of my friends took out a bottle of wine from his bosom and suggested we drink it in our close circle. So, how could we be apart from the whole company? Then our eyes glanced on the ladder which led to the flat roof of the house. In a minute we were sitting on the roof and were drinking with laughter, happy with ourselves and all the surrounding. It was a crisp-snow Russian winter night, with cold bright stars twinkling in the sky. But because of that laughter one of our friends came out of the house and, having seen such a picture, took the ladder away - saying, "Sit there for a while, get a little cold!" In ten minutes our wine was over and we really got cold. And then, looking down, I cried, "It's only three meters here, nothing for a jump! It's our usual practice during trainings!" And jumped down without any hesitation...

Indeed, I had specially trained in jumping from heights for mountaineering and had reached certain deftness in that, but either the coordination was not so good after the night walks, or the height was different, but when I touched down I heard a loud crunch and stayed in the snow with an unnaturally twisted leg...

Such a moment is a bifurcation point, the point in which one might stop, look back, and think about everything that has happened and what could happen. At such moments there is a possibility to start something new. Such moments take place in the life of any person, but not all make use of them. There are obviously some higher forces that organize these special moments. Nothing happens for no particular reason. Nothing takes place without some intrinsic meaning. Moreover, some people can catch this intrinsic meaning, others disregard it and continue to run further, with the eyes closed and the bit taken between the teeth, as a horse in harness.

As a result of my jump I was tied to my plaster leg cast for half a year and instead of an active working bustle, I was absorbed in the slow inertia of books, thoughts and friendly chats.

And so, reading books on Eastern philosophy, Buddhism, and especially Zen Buddhism, I suddenly realized that Fortune gave me a rare chance to absolutely officially start a unique activity, completely unknown and full of mysteries and enigmas! And I still wondered if I should start it or not! If I had lost that chance, another would not have come so soon! And I should consider everything done before as good schooling, which gave me practice and certain skills! When I fully realized this thought, I called my boss and told him,

"Dr. Ban'kovsky, could you come to me to discuss one important idea?"

(Being tied to my leg cast, I could only receive guests myself). Hardly had the boss entered the room and started an intellectual conversation about weather and rehabilitation measures for the broken extremities, as I struck him with the announcement,

"Dr. Ban'kovsky, I decided to leave the topic of electron emission and fully dedicate my work to the Kirlian effect."

That was a complete surprise for the boss. He tried to explain to me how premature my decision was, how many difficulties it would entail: "to close" one topic, to prove another one, to rearrange the post-graduate studies, and finally start working from scratch. A huge amount of work had been performed on the previous physical topic, a complex setup had been designed, and several articles had been published. In principle, everything was ready for the defense of the thesis in the nearest future. Nothing was done on the new topic. Only ideas, some results and absolutely uncertain relation from the scientific establishment. But I was steadfast. Dr. Ban'kovsky asked me to think again and left, distressed. In half a year, when the leg turned from a burden into a support, I fully began studying the Kirlian effect.

By the way, interesting twists of Fortune: in three years, when I was finally getting ready to defend my thesis on physical processes of the Kirlian effect, a book was published in Holland. The processes of the two-component absorption, which I had once been working on, were studied there perfectly and in thorough detail. What was more, all the processes were studied on such an instrumental level that we couldn't have even dreamt about at that time. And I imagined how I would have looked, having finished the work, having prepared the thesis, and then having got the book where everything had already been studied in detail.

Some time later I met a girl who was in a similar situation: she was preparing a dissertation on the redirection of the Siberian rivers into the Middle Asia, the work was already fully done, when under the pressure of sensible ecologists the project was closed. So was her dissertation. By the way, imagine what political collisions could come out in our time of such a project, if it were put into practice! In the Soviet Union the thesis topics were divided into two big categories: the first one - topics connected with the military complex. Funds for these topics were allocated, they were included in the plans, articles were published, and everything had actively been promoted. And other topics, which practically didn't exist. Maybe, somewhere in the social sciences such topics were considered. But rather not, as the social sciences had always been mostly politicized. And even philology should have proved the advantages of the Soviet system and socialistic realism over all other systems and ideologies.

Therefore, one way or another, the entire powerful Soviet science worked for the military machine. All the topics developed at scientific institutions were to be strictly controlled, selected, and registered in plans. And the plans were always checked by the "scientific officers."

Now I understand what a pain in the neck was my decision to the administration. On the one hand, the new topic was initiated by the higher authorities; on the other hand, it was in no way connected with the military complex.

But my decision was firm, so, after a couple of weeks of persuasion, I was allowed to study the Kirlian effect.

And after some time another bifurcation point appeared. The society of naturalists named after professor Popov was founded in the city. An association of people who gathered all together once a month and listened to the reports on new unusual approaches in science was hiding under this innocent pseudo-scientific name. First of all, they discussed approaches connected with the influence of the human psyche on the environment.

In Moscow, professor E. Kogan managed to organize active work in such section: seminars with the participation of famous scientists, meetings with foreign specialists in parapsychology, and experimental investigation in a small laboratory in the center of Moscow. Something similar was planned also for Leningrad. Meetings and lectures took place in one of the largest Leningrad halls ? the Palace of Culture of Lensovet, and gathered a lot of people. Here I heard about Wang and Philippine healers, Uri Geller and telepathic experiments for the first time. Each lecture aroused active discussion, but there were not less furious critics than enthusiastic adherents. Attempts to organize experimental work were undertaken several times; what was more, the Kirlian effect occupied an honorable place within the list of other perspective methods, but nothing of these plans was realized. Finally, the entire work of the section was closed ? a decision was taken by the ideological department of Central Committee of the C.P.S.U. that parapsychology does not entirely correspond to the Marx-Lenin directions, and a military campaign was started. Abusively mocking articles of a biophysicist Volkenschtein and a story-teller L'vov were published in the central newspapers. Words "charlatan" and "deceiver" were the most obscene in these articles, and the Kirlians were called "an idle couple from Krasnodar who waste everyone's time with their harmful photographic "hocus-pocuses". Naturally, the activity of the bioenergetic section was closed by a higher secret resolution.

Now it is clear that it wasn't just a mere interest of people to the new scientific phenomena. That was an attempt for a spiritual outcome, beyond the bounds of strict censorial limits and the materialistic paradigm in which the Soviet people were held. If it was impossible to make a step in another direction without punishment, if it was impossible to go out of a lifelong socialistic prison, it was possible to do something apolitical having no connection with real life: climb vertical rocks and dream about something abstract. And, as soon as every person strives for something unusual, high and spiritual, all attempts to suppress such a striving lead to inner uprising. And any possibility to make something for one's own soul results in an inner rise and inner festival. Thus it was with the first Christians, who were bringing a sparkle of their spiritual faith through all the persecutions. Thus it was during the periods of most cruel religious oppressions. All these were the movements of soul, the soul thirsting for renovation and spiritual development.

In contrast to the majority of animals, apart from physical food a human being needs spiritual food. This is that very powerful stimulus which furthers nations, causes the development of civilization, and creates the spiral of history. We live not in the world of material things and formations, but in the world of powerful spiritual movements.

The Spirit is the moving force of History.

But even that small period of time when all this existed gave me an opportunity to meet many interesting people, including a whole series of capable extrasensory individuals. I don't want to specify all of them by names and, what is more, give any characteristics. Some of them have become world-renowned, have published books and have often performed on television. Others have disappeared from the scene and nothing is heard about them now. Naming some and not mentioning the others, one could be offended, and another one could have extra advertisement. I wouldn't like to do either the former or the latter. During many years I was moving in the circle of sensitives, and not just moving, but professionally studying this phenomenon. I participated in seminars, workshops and training camps. Books, meetings, communication ? all these added a fragment in the mosaic of a general picture. I have gone through most of the techniques, ? have wanted to feel the essence of that, to make sure of the reality of the vague mystical phenomena. The fact is that each person wants to try everything himself or herself, to touch and to feel. Of course, in general this pertains to the things in relation to which the public opinion is not formed ? nobody really doubts the existence of Galaxies and Black Holes, although no one has ever managed to see or to touch them. Modern concepts of parapsychological phenomena will be discussed in the next chapters of this book, and now I would like to tell about one bright, talented person, whose fate was very typical of the whole endeavor to understand mystical phenomena.

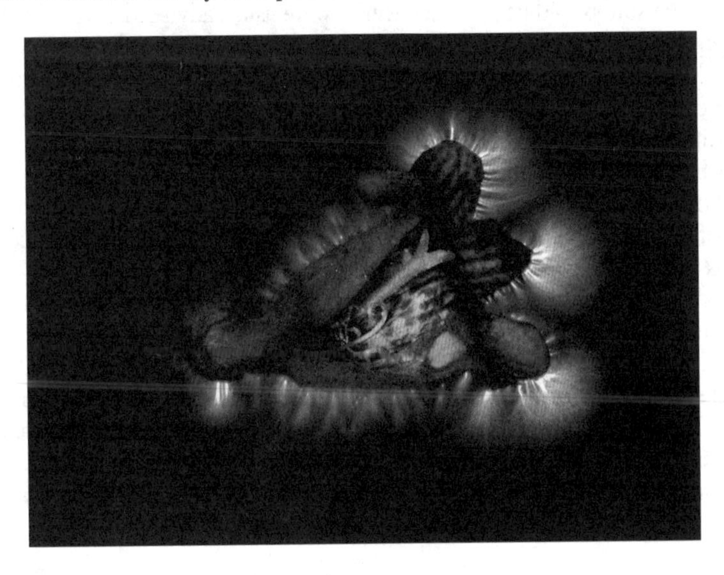

EXPERIMENT

At one of the meetings of the society of naturalists I made acquaintance with Vadim Polyakov. He was a teacher at the air force college at that time. He was a very active, thin young man, making a successful military career. The only rub was that he possessed a very powerful talent for healing. His talent appeared purely by chance, when he managed to cure his wife and then several colleagues from headache and radiculitis using his hands. He started developing these capabilities, and he began acquiring and reading the rare books on yoga. Once someone confidentially brought him a "self-published" book on Theosophy by E. Blavatskaya. At that time such literature was illegal in the Soviet Union. One wouldn't really go to prison for reading it, but a person's troubles could be great. Moreover, there was a circle of people who were getting these books, translating and publishing themselves. The next book came after the first one, the next meeting after the previous. Vadim reached a high level of self-development. And being a common teacher of physics, he simultaneously started investigating paranormal phenomena on his own initiative.

After the first acquaintance with him we met several times in various gatherings, and we liked each other. It was a usual non-committal acquaintance until finally the next bifurcation point appeared. My wife was in poor health and often had one or another disease. It was a natural reaction to the problems of life: a permanent lack of money, routine work, and two unhealthy children. During one of the visits to the doctor, the UHF-therapy was done, which was a common procedure at that time. In a couple of hours the lower part of her stomach was aching awfully. The doctors whom we consulted told us that such things were usual, advised her to take Analgin (an analgesic drug), and wait until everything passed. Several weeks went by, no procedures helped, she was suffering, and we could do nothing. Then an idea came to my mind: I called Vadim and asked,

"Vadim, maybe you can help?"

"With pleasure", he replied.

And one evening, after work, he came to visit us. We had dinner and drank half a bottle of cognac. After that he gave a back massage to my wife, and in a couple of minutes all the pain was gone.

"This is an energy massage", he explained, "I learned it from one of the most famous masseurs of Russia, who was working at the Kirov (Mariinsky) theater. He helped the dancers to remove pain. It is said that the people of quality went to him. He was several times invited to work in the hospital for the privileged communist figures. He always refused, referring to the fact that his occupation was ballet. Obviously, he enjoyed communicating with young, beautiful girls and young men. This brought great pleasure for his soul, and he didn't want to pass to fat communist bosses and their wives, notwithstanding the high salary.

Vadim was assisting this person for several years, and the masseur passed to him all the secrets of his magic.

That occasion surprised me so much, that I started believing in the reality of extrasensory phenomena at once. This subject captured my imagination and became a topic of scientific research for many years.

After a 15-minute procedure performed by Polyakov, my wife started flying around, chirping and preparing appetizers for us. Discussing the situation, we finished drinking the cognac and I asked Vadim,

"Perhaps, this way you can influence not only people, but physical devices?"

"Of course", replied Vadim, "I can influence anything, the whole world is unified and identical in its essence. If you know how to influence it, you can work both with people, animals, and non-material objects. They are all subject to the same laws."

"Let us try! We can make some experiments".

"OK. Think what might be done, and we will organize it together".

Participating in the meetings of the society of naturalists, I came to the idea that it was necessary to try various physical systems to show the influence of human consciousness on the physical world. If people can influence each other, can communicate with animals, they can influence physical processes of the outer world under certain conditions. This is said in all the fairy tales and legends. And as a Russian proverb goes, "Just look at the galouschka and it's in your mouth". (Galouschka - a dumpling, a small ball of dough, an Ukranian dish).

An idea came to me that such a structure can be represented by a system which exists in various phases, i.e. in several stable states, so that a transition from one state to another could take place under the influence of a weak effect. And these states should significantly differ in parameters, so that the transition could be easily observed. Imagine that a caterpillar crawls up a sloping branch slowly. If you look at it, it's very difficult to observe that it's moving, because it slowly crawls over from one place to another. The progress will be seen only in a long period of time. At the same time, if a caterpillar falls down from one branch to another, you will see it at once. Thus, there are smooth processes, and it is very difficult to observe how a smooth process changes. And there are spasmodic processes, when it is easy to see the transition from one state to another or from one phase to another.

After some meditation I built a sensor from materials at hand. It's worth mentioning that materials at hand were rather exotic in our laboratory: tungsten and chrome plates, molybdenic wires, quartz plates and vessels. It presented no difficulty to make any construction of glass, because we had a good glass blowing shop with wonderful masters. And so, after some time, I made such a construction, placed it in a special apparatus, and tested it. I made sure that under stable voltage and environmental conditions that construction worked for a long period of time, maintaining the same output parameters. And if voltage changed slightly, it jumped spasmodically from one state to another or from one phase to another. Having spent the day experimenting (carrying out the experiments), I called Vadim and invited him to visit the laboratory one evening, to make the experiments.

He came the next evening. No one of the employees was in the laboratory, only the person on duty on the floor - we didn't want to inform anybody about such strange experiments - they would have only laughed. I turned on the apparatus, explained to Vadim the principles of its operation, and asked him to try to switch its mode from one phase to another, by the force of will. At the same time I was observing the number of impulses corresponding to one or another phase on the screen of an oscillograph. Vadim stood at a distance of two meters from the device. He concentrated. His face became thoughtful, meditative. Later I saw him several times like that when he was working with patients.

A sequence of jumping impulses of current ran across the screen of the oscillograph. They were like tiny soldiers - some were higher, some lower, but on the whole approximately the same. And suddenly, after a time, the current pulses started jittering and spasmodically increasing in size and quantity.

"So, how is it?" asked Vadim after a while.

"Seems something is going on", I replied.

"Well, let's have a rest", Vadim relaxed, smiled, and after a second the impulses came back to the initial state.

We repeated the experiment. The result was the same. The device reproducibly changed its readings under the influence of mental concentration performed by Polyakov. It was a total shock for me. Truthfully speaking, preparing the experiment I had never believed that something could really come out of it. It was far beyond the laws of our materialistic physics, to which I was accustomed. But everything worked, and I was watching the change of phase of the device's condition with my own eyes.

Then we drank some tea and repeated the experiment again. And again the result was positive.

"If I hadn't seen this with my own eyes, I would never have believed", I confessed to Vadim. He smiled,

"Yes. The majority of people play the role of the doubting Thomas. They should touch everything themselves".

"Then, let's record all that we've done".

" OK", agreed Vadim.

The next day I prepared the equipment for recording data from the oscillograph. Vadim came to the laboratory at 7 p.m., and we were recording and photographing until midnight. Eight of ten attempts of Vadim's influence came out to be successful. The last two took place very late; Vadim was tired, and, perhaps, that's why there was no result. Thus, we supposed that we gathered absolutely objective, irrefutable proof of an effect of human consciousness on a physical system. We repeated that result in another three days. And, again, with positive effect. After that I took the results of all the experiments, described them, prepared materials and solemnly put them on the boss's table.

I can't say that he was especially enthusiastic about that. After long hesitation, he offered to appoint a commission and carry out the experiments again. Then a series of consultations followed, but nobody showed much enthusiasm and nobody was in a hurry to do anything. That surprised me greatly. It seemed that we could create a new phenomenon of the outer world, prove the influence of Consciousness on Matter. Wasn't it interesting? Why did we hesitate?

Only many years later did I realize the reason for such an attitude. Indeed, these experiments strongly undermined the basis of materialistic science. And in the beginning of the 1980's Soviet masters began to understand that. Just then a tense struggle for maintenance of the Ssoviet system was under way. The struggle was waged first of all in the sphere of consciousness and ideology. A powerful movement of dissidents was arising, and thousands of people became involved. They were fought against, put into prisons, concentration camps, and sent out of the country by many means. Then many people emigrated to Israel and the USA. Anyone who caused some doubts would likely be thrown out of the Soviet Union. That was the epoch of legal processes against dissidents. The whole country was disputing about the traitor Sakharov, without any idea of what he knew and what he was talking about. Solzhenizin, Brodsky, Shemyakin, and hundreds of cultural workers were sent out of the country. The church was officially permitted, but in fact it was suppressed and persecuted.

Therefore, experiments on the influence of consciousness and spirit on the materialistic world represented a real threat for the Soviet system.

The world should be material! Everything in the world is based on the movement of material and money. Human senses and human emotions have nothing to do with this world. Love serves for reproduction. The Soviet masters did their best to support that philosophy, and, being clever and experienced, they fully recognized the danger of any movements, even indirectly proving the role of Consciousness and Soul in our world.

From the very beginning the science studying extrasensory abilities was a spiritual, idealistic movement. It radically undermined the basics of the materialistic theory of Marxism-Leninism. It maintained that apart from the physical cover of the material world, there was something else. Something unknown. But the Soviet life was based on the idea that the authorities knew everything, and there were no secrets from them. The opposition to paranormal studies proceeded in several ways. The public, official way was that the most prestige and popular newspapers published slashing articles by the leading professors, where it was claimed that "it can't be true because it can never be so" or "it's known to science and contradicts its laws." The famous biophysicist Volkenstein, a great professional in his sphere, was especially notable among such authors. He had authority, wrote good books on traditional biophysics, and at the same time strongly disputed everything new and unknown.

Another method of opposition to paranormal studies was the direct prohibition of any experiments in this sphere. Special secret directives were circulated, which were intended only for the leading employees. One of those directives categorically prohibited any work on the investigation of unusual psychic phenomena. Such experiments were carried out solely in secret laboratories of the KGB.

How could I have known all that at the time? I was far from any political ideas and was always standing aside from dissident movements. The only things which had really fascinated me were science, new knowledge, alpinism, and... pretty girls. So, I had to leave aside the experiments with Polyakov; moreover, he lost much interest in that himself. Being a person who could be easily carried away, he supposed that everything had already been proved and it was simply boring to repeat the same things time and again. In addition, the defense of my dissertation was coming close. Now on the Kirlian effect. This required much effort and time. Night after night I was clearing a place on the kitchen table, making strong tea, and writing one page after another.

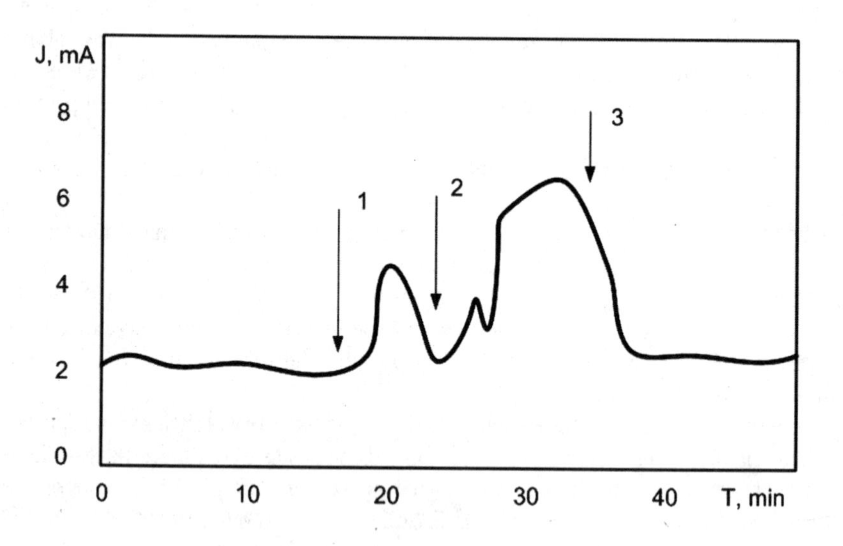

Change of Sensor signal under mental influence.
Moment 1 - beginning of mental influence from the distance, moment 2 - next influence; moment 3 - the end of influence.

END OF CHAPTER

The dissertation was accomplished. I presented it in a usual physical context, not mentioning any special phenomena. All the work was dedicated to the analysis of physical processes of the phenomenon: physics of discharges, movement of electrons, and experiments under various conditions 8.

And the defense was very successful. Both in the Soviet Union and modern Russia the defense of a dissertation is a complex and serious process. About 15 scientists working in this sphere participate in the defense. A candidate for the degree presents the work written in hundreds of pages. Opponents study it and make their judgments. And, despite the large work, some defenses do not end with awarding a degree. There are cases when the Academic council rejects the presented work and sends it for revision. There is also the Higher certification commission, which studies the presented documents and estimates if the given work meets the scientific level. Of course, there are many nuances in any system, and they have been inevitable in this system, too. Particularly, contacts and connections have played an important role. A special question was the defense of dissertations by people who had been quite far from the scientific process and had been moving along the party line. And in the sphere of such sciences as social science and economics, dissertations were mainly a political event, not a scientific one. This influence pertained to physics to a lesser extent, although even here a director of some institute, a former party worker could always count on a positive attitude to his dissertation. At the same time this system made people to be earnest about the work, and many wonderful investigations were carried out because their authors aimed at making a good dissertation. As it is said in a common proverb, "A normal scientific work has to be done at least once in life before the dissertation." And so my dissertation on the Kirlian effect was discussed, accepted and highly evaluated. I obtained the diploma, which stated that the rank of a candidate of science was awarded to me, equivalent to the rank of Ph.D. in the Western system. It was a fulfillment, a conclusion of a life stage, an ending of a chapter in life. I felt that new horizons were opening before me. And that summer, there was a situation which showed again that there are some forces directing our life and our fate. There are several peaks higher than seven-thousand meters in the Soviet Union which are a Mecca for all alpinists. To climb a seven-thousand meter peak meant to get into the next category of the alpinists' elite. There was even a special name - "snow leopard," which was given to those who managed to ascend all four Soviet seven-thousand meter peaks. Of course, perfect physical training and special physical and moral qualities were required. In addition, it was necessary to belong to a certain system, as all trips to the mountains in the Soviet Union were paid for by certain trade-union organizations, and not by individual climbers. It was impossible to go and start the climb outright. There was a powerful system of control which checked everybody who was climbing high mountains. The system served for security, and it provided a quite low fatality rate. But at the same time it was a means of control. Another control system, created by the genius Jesuitical mind of Joseph Stalin!

An expedition to one of the seven-thousand meter peaks - the Peak of Lenin - was announced. That was an anniversary expedition; many people wanted to participate in it. Everything was arranged with great pomp. It was considered to be a high honor to participate in this expedition. I was invited to take part. But before that I had already promised to be in an expedition to the mountains of the Pamirs to climb one of the most technically complex peaks. I had to make a decision: either to participate in the climb to a seven-thousand meter peak and reject the wall ascension, and thus put the team in a difficult situation, or try to somehow combine both events.

After some consideration, I decided that if I had promised I would keep to my word. What was more, the tradition of bringing up young alpinists was strong at that time. That was considered to be everyone's sacrosanct obligation. Therefore, I decided to first go to the Pamirs and work with young guys for 20 days, and then, if everything was fine, go to the seven-thousand meter peak and try to take part in the anniversary climb.

After 20 days spent in the Pamirs and after a successful climb I came back to Leningrad to spend some days at home and then go to the seven-thousand meter peak. But terrible news came. Everyone who was at the peak of Lenin, almost everyone who participated in this expedition was gone.

The situation seemed to be absolutely improbable. Over many decades the rise to this peak had gone through one and the same known route. The first time, in 1930's, the expedition under the direction of Kirienko had followed that route, from the base camp on the lower glade to the great side at the height of more than 5 kilometers (16,500 feet), where the first camp had been set. The place had been absolutely calm and safe - a large icy plateau, stretching for several kilometers. From there a steep icy slope leading out practically to the top could be seen. Starting from the plateau early in the morning, one could have risen to the top and come back before night. Each summer for decades the alpinists had come to this plateau, set up their tents and started the rise from there.

And so that year dozens of expeditions gathered on the plateau - more than a hundred people. This year the weather was unusual, even for the Pamirs. Each day it was snowing, and alpinists had to make their trails again and again in deep snowdrifts. As a friend of mine, one of the six people who had survived, told me later, "When we came up to this plateau by night we saw that all the tents were covered with snow, only the tops of tents were seen".

People were coming up to the plateau during the whole day, putting up their tents, and it was snowing and snowing. The alpinists were getting into their shelters covered with snow, but even cozier so, they were setting primus stoves, drinking tea, telling anecdotes to one another, and relaxing until morning, in order to start the next stage of the climb before dawn. By night the whole camp was fully covered with snow, so it was even difficult to guess that there was a whole camp under the white cover.

Lesha and his friend put their tent above the snow thus they were above everybody else. They came to the closest tent's top seen from under the snow and dug a trench to the entrance with a snow shovel. The familiar faces showed sluggishly, swollen up from long hours of lying in sleeping-bags. They greeted one another and discussed the situation. Everything appeared calm and was proceeding in accordance with the plan. Having said good-bye until the next morning, the guys got into their tent, drank tea, got into their sleeping-bags, and, having put their boots under their heads, went to a calm, youthful sleep.

At night they woke up from a terrible crash. It seemed that the air itself was thundering. Lesha started looking for an electric torch to see what was going on. And the moment when he took the torch in his hand a powerful wave caught them, threw them into the air and carried them. Carried them away, rolling, raising into the air and throwing them into the snow. Everything mixed up in a wild crash, howl, and roar. Thus it continued literally for many tens of seconds. Then everything calmed down. Full silence fell.

Lesha came to his senses, opened his eyes and saw an absolutely clear starry sky above. Total silence; he was lying almost fully covered with snow. He started scrambling out, brokenly rowing with hands and legs, got out of the snow, and only then he felt the crystal frosty air of the mountain night. Having looked at himself, Lesha saw that he was standing in his track-suite, without boots, without hat, without jacket. When he looked around he saw an absolutely flat snow desert. He took to his heels, fell in the snow waist-deep and stumbled over something firm. He started digging brokenly, and in several minutes found a human hand. Feverishly digging the snow he dug up his friend, who was lying one meter under the snow. Lesha pulled him out, shook him; he slowly opened his eyes and asked, "What's happened? Where are we?" When the friend regained consciousness, they started to find out what had happened.

An event which couldn't have been predicted by anyone took place. That year was very snowy. The snow fell for days, and great snow beds gathered in the mountains. And so, either under the pressure of the snow bed, or because of a small earthquake, high up in the mountains, at a distance of more than two kilometers from the plateau, a great snow avalanche fell down. It flowed down, picking up speed, passed a wide icy crack, rolled out to the plain of the glacier, and whirled along that plain in a two-hundred meter wave, carrying all before it.

That avalanche fully covered the camp, where dozens of alpinists took (had taken) their rest. And the air wave coming before the avalanche swept away a lonely tent standing above, and threw the two guys out on the surface, who were the only live witnesses of the dreadful event.

When they realized what had happened, of course, not entirely, only in a general way, they rushed digging out the place of the camp barehanded, in track-suits and socks as they were. But however hard they dug, they couldn't find any trace of their friends. Afterwards several expeditions with spades, with a large stock of food came up there, but after a week of persistent excavations the only things which were found were a few objects. Thus the snow desert turned out to be the grave of 70 people. And nobody will ever know what feelings, what horrors went through those guys slowly suffocating under the snow.

The only team which survived that tragedy was the team of Vladimir Baliberdin, an outstanding sportsman, one of the strongest alpinists of the world. He was one of the first alpinists who could climb without oxygen to the highest peak of the world - Chomolungma (the Mother Goddess of the World) which the English call the Mount Everest. Undoubtedly, he was the strongest alpinist of the Soviet Union, he had an amazing feeling of the mountains and a surprising self-confidence. Therefore, the night before the avalanche he suddenly roused his team (three people) and said,

"Let's go up!"

They tried to talk him out of it,

"You are crazy! Who goes up at 7 p.m.! What for? Spend the night, and go tomorrow at 4 a.m."

"No, we should work. We should train camping at night, so let's get up and go right now".

Trying to dissuade Vladimir was useless. If he had made some decision, he didn't listen to anyone. He had inner feeling and sense. This had always caused him a lot of problems with friends. And he had practically no friends, but at the same time it made him the greatest alpinist of the Soviet Union of that time, and, maybe, of the whole world.

He raised his team. They put on heavy boots, helmets, took equipment, rucksacks, and left at night. They went 200 meters up, but it was enough to hear the rumble of the avalanche from the above. In an hour they came down and tried to provide help, but it was impossible to find anyone then.

I've already written about the laws of life and death in my book "Light after Life." One of these laws says that it is very dangerous to be always successful. Great success leads to great catastrophes. Everything has to be paid for. It is better to have small problems periodically. Then they protect from big problems. This is clearly seen in practically every step in alpinism.

Vladimir Baliberdin, an outstanding alpinist, he ascended the most difficult peaks solo - on his own, took part in the Himalaya expeditions as a leader. An alpinist who had never suffered a defeat in the mountains. He possessed unique alpinist's skills. He had always succeeded in what he had intended to do. He had always moved ahead. And this very person, risking his life every season and always with great success, passed away in an absolutely absurd way. His car was crushed by a huge truck which was driving at high speed against the light at a crossing. A drunk Finnish guy was driving.

TELEPATHY

The middle of the 1980's got the name of "stagnation" or "depression." The old Brezhnev was no longer controlling the situation in the country, but every state official was trying to do his best to maintain that situation without change. Officials and old people are afraid of changes. Everyone understood that something had to be changed, but nobody could make up his mind to begin. The group of old politicians headed by Brezhnev and the KGB chief Andronov held tenacious control of the mechanisms of power and the state machinery. The system created by Stalin was so stable that it outlived its creator for almost 40 years. A unique case in the history of dictatorships!

At the same time it was a good time for the Soviet science. The governors of all countries accepted the practical power of science and started treating its workers with respect. And even the most extraordinary scientists, such as Einstein or Bohr sometimes got very practical ideas. The society had to put up with their extravagance and listen to their ideas, even if those appeared crazy at first sight. Especially if those ideas promised some advantages for the military sphere. Money was not spared in that case.

Therefore, in the 1980's several projects on using psychic energy for military purposes were started in various countries. As Fate willed, I happened to be involved in one of such projects.

The task was purely practical: to create a system of telepathic communication with submarines. Radio communication with deep-sea submarines is a very complex task. The problem has been that the radio waves almost do not penetrate in salty water. So the submarines to surface in order to communicate with the chiefs. Obviously, new physical principles have been needed for the communication. Here one of the big bosses read a popular book about paranormal abilities came across a chapter on telepathy and found that it had been successfully used in the world. After a series of meetings of higher authorities a decision was made to organize a laboratory and entrust it to develop a system of telepathic communication within three years.

I got into this story by chance - meeting a friend during an out-of-town walk, who, as it was found, had been commissioned to create that laboratory. We hadn't seen each other for five years. After exchanging banalities, he suddenly asked,

"I've heard you're studying the Kirlian effect. Is it true that psycho-emotional states can be registered by this method?"

"Of course, this is one of the main advantages of the method", I replied. In a week (obviously, after thoroughly checking my reliability) I was offered the post of deputy chief of the laboratory with quite a good salary".

Thus, a meeting by chance led to the next turn of fortune. But was that chance accidental? Many things have their inner secret laws. Interestingly that this meeting took place in the same village, Solnechnoye near St. Petersburg, where I had once jumped from the roof so "successfully." I came to this village for the first time after a long period....

Working conditions in the laboratory were almost perfect. The resources were practically unlimited. Any equipment was given to us at our option. The epoch of PCs was just beginning, but our software team obtained all the latest equipment. Of course, from the today's viewpoint the latter seems to be toys for children.

But the most splendid advantage was that we were given full freedom of action. We could do what we considered necessary and were only to periodically report on our results. Of course, there was no way to avoid that requirement.

No method of mental information transmission developed in the world provided safety in data transfer. And that was one of the main requirements for practical application. From the viewpoint of the information theory any communication channel consists of the three main elements: the transmitted message is encoded, comes to the transmitter (sender), is transmitted to the receiver through the communication channel, and then decoded. For example, visual information is encoded into a TV signal in the television channel, analog or digital; radiated in the form of radio wave by the transmitter; comes to the receiver tuned to that wave; and is decoded into an image in the TV set.

In the case of a telepathic channel of communication, the receiver and the sender were persons, we could only assume that the communication channel existed, but the process of encoding/decoding needed to be thought through. What could be sent? The Zener cards? Pictures? Emotional images like views of a calm sea or horrors of war?

After a few weeks of reflection and debate someone suddenly recalled healers.

"What do they do in the course of treatment?"

"Influence other people."

"Can the moment of this influence be measured?"

Books and magazines were looked through. A few publications on transformation of ElectroEncephaloGrams (EEG) of patients during healing sessions were discovered in some exotic magazines. Reliable changes were obtained.

"And what about hypnosis? Also a very similar process".

Again a search in catalogues. The catch was poor, but rather encouraging. It became obvious that in the process of mental influence the electrical signals of brain changed.

"And what happens with the Kirlian glow? Will it change?"

Nobody knew. We found nothing in the literature, but it was worth trying. Thus, the scheme of an experiment was born. The transmitter-inductor got a signal (a lamp flashed) and sent the influence to the receiver. The latter was sitting in the laboratory, in another room, and EEG and Kirlian measurements were constantly being taken from him. At the first stage we looked to see if changes of electrical signals took place, synchronous changes following after the blinking of the inductor's lamp.

Another question: should we take random people for the experiments, as the

Americans do, or should we select specially? Here we unanimously decided: special pairs should be prepared. The practice of healing has shown that there are talents, Masters and apprentices. Anyone can learn to feel energies, but one should be gifted from God in order to work really effectively.

We contacted a colleague who was managing a large center on training psychic abilities, and started the experiments in a few days. To our surprise, the result was obtained in the very first experiment. The healer, a pleasant 40-year-old man, brought a patient with him - a pretty young woman. We explained the aims of the work, demonstrated all the equipment, and they agreed to participate in the experiments with pleasure. They were interested to see what would the result be, and if we would manage to observe something.

It is especially worth mentioning that it is very pleasant to work with healers and extrasensory individuals. As a rule, they are willing to participate in the experiments and are interested to discuss the results. Although they are psychologically very different, as in any group of people, and often a certain approach is required to work with them. But let's discuss that later, in the next chapters.

So, the inductor and the recipient were sitting in different rooms, the walls were covered with wires, and in 15 minutes of relaxation exercises we gave a sign to the inductor, "Go!". The recipient was relaxed, but keeping awake. We decided not to use the dream technique, although it was also very interesting and revealing. A lamp flashed and the inductor's face got a concentrated look. In a minute he visibly relaxed. Again a flash of lamp, and then a stop. Thus, in the course of time, we learned to transmit the signals of the Morse code, i.e. complete phrases and messages. On completion of the experiment, the data taken from the recipient was processed and the moments of change in the signal were found by EEG and Kirlian. These moments were marked on the time axis and were compared with the moments when the lamp flashed. Victory! Reliable coincidence in almost 70% of cases! The communication channel worked!

We spent nearly a year after that working out the details of such experiments. We had checked several dozens of pairs, and revealed an interesting pattern: even the best pairs lost the effectiveness of work in the course of the experiments. The first and second sessions were the most effective; then the effectiveness reduced, reaching a stable level by the fifth or sixth experiment, and further remained on that level. The frequency of spontaneous, nonprogrammable telepathic contacts between the members of pairs increased conversely with the decrease of effectiveness of directed transmission or directed influence. The persons felt the moment when it was necessary to call one another, they saw the same dreams, they bought the same clothes, products, and magazines simultaneously and independently, they were sad and happy at the same time. Moreover, that didn't depend on the age or sex of the people in the pair. The only requirement was mutual sympathy. (Fig. 47)

A year after we were able to report 85% of effectiveness of the telepathic communication channel. The system was working. There were methods of selection and

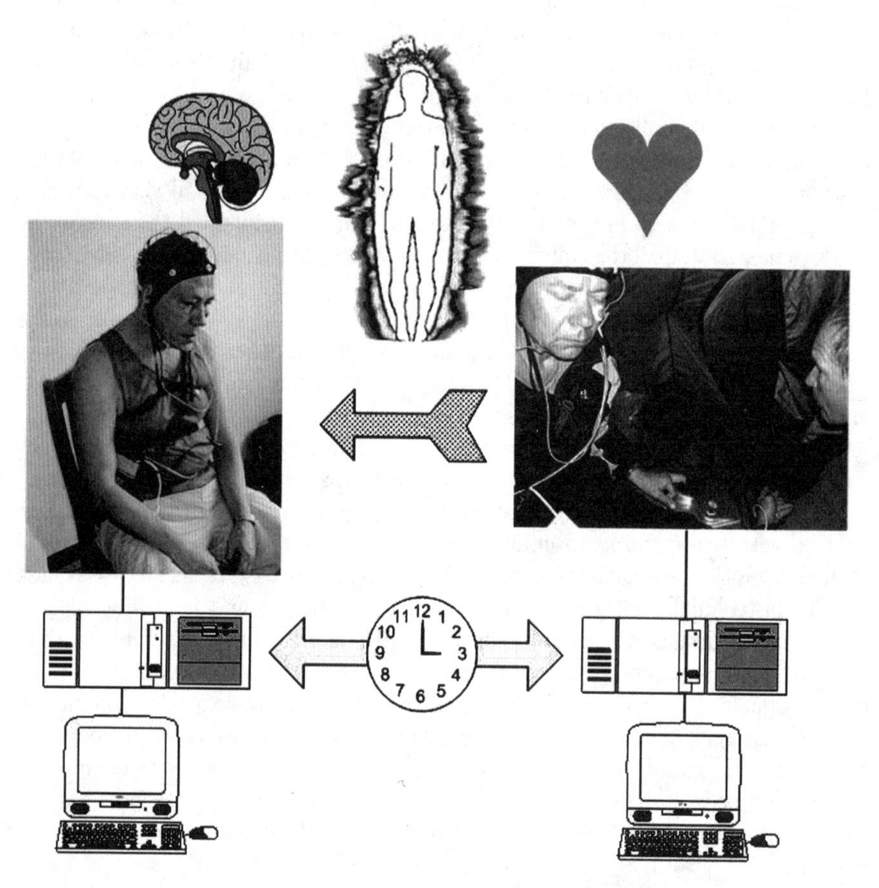

Fig.47. The principle of telepathy experiments.

training of the participants of experiments, conditions for transition into the altered state of consciousness were trained, which was necessary for the effective work performed by the inductor. Naturally, we had undergone all the trainings ourselves and after some time we introduced telepathic communication into our everyday practice. At quite a simple level, of course. Even today when I need to contact some person and the situation can allow for a couple of days, I concentrate on the image of this person and ask them to contact me. After some time I receive a telephone call or, what is even stranger, meet this person in the street or in the metro. A very convenient method: saves a lot of time and money, especially on international calls.

After many years the life spiral made another convolution, and I became involved in the investigation of telepathy again. It took place during our expeditions to the Himalayas, to the foot of Everest, and to Canaima mountains in Venezuela, from where our Moscow colleagues organized sessions of mental communication with Moscow, and we measured their energy state with the help of our techniques. But that is another story.

At the same time, in the 1980's we came to a conclusion that telepathy wasn't appropriate for the communication with submarines. 85% of effectiveness would have been quite acceptable. It was always possible to increase the reliability of communication by way of parallel channels. But the human factor played rather a significant role in this process; the factor of the percipient.

In the process of telepathic communication one of the participants was active: he was supposed to send information; the other one at the agreed time of the experiment was lying in the state of calm relaxation. And, it turned out that it was very difficult to maintain this state of doing nothing within a long period of time. Especially sitting inside a submerged submarine. The person soon would become bored. He would either start to fall asleep during the experiments, or lost his sensitivity, and began to think about getting out. This wasn't suitable for military purposes. Some decisions had to be searched for, or the program had to be closed. Nobody wanted the latter - everything came out to be very interesting and exciting.

SENSORS

At some moment we came to a unanimous decision that it was necessary to work through human influence on physical sensors. Keeping a percipient on a submarine during several months was expensive and not gratifying work. But if we could mount a sensor sensor, reacting to mental influence, the channel of communication could become practically useful. I recalled our experiments with Vadim Polyakov and made drawings of the sensor in half a day. In a month the experimental setting was ready.

By that time we had gathered working experience with extrasensory individuals and had come to understand that this work required special conditions. In the city, a person came to the experiments for a couple of hours, tearing himself away from other business, and some time was required in order to get away from current things and come into an altered state of consciousness. This state could be maintained for quite a long period of time, but the working day ended soon, the night came, and the experiments had to be stopped. Therefore, we made a decision to organize a summer session, out of town. As I already mentioned, money was spent without limitations during that time. We wrote a paper to the administration and explained everything, and after some time were already searching for a summer lodge. Having looked through several variants, we settled on a building of a medical college in the Baltic region. There were no studies in summer, and we managed to rent the building for two months. Conditions were perfect: a small Baltic town, having kept its national style and fascination through 40 years of the Soviet occupation: cozy houses under tiled roofs; a Roman-Catholic church, that survived the Bolsheviks' devastation due to heroic efforts of local patriots; and a huge tennis park. By that time the population of Pribaltica had already submitted to their position within the Soviet empire, the Russians were treated friendly, and only in the circle of close friends during heart-to-heart talks could one feel a secret hatred to the invader's country. In the 1980's we felt comfortable and at ease in Pribaltica.

We brought trucks full of equipment, and in a month everything was ready for work. Half of the rooms were prepared for experiments, the other half for habitation. The building was in the heart of the park, surrounded by high oaks, and our activity didn't attract much outside attention.

We selected a group of talented sensitives, following the results of experiments carried out the previous year. They lived and worked in the summer center for two weeks, leaving it just to walk in the park or visit the nearest store. In a couple of days the person already forgot about his business and troubles, and was fully absorbed in the atmosphere of meditation, measurements, and leisurely evening talks. Enthusiasts had an opportunity to go to parties in the nearby health center, where groups of girls and women didn't mind persistent signs of attention of the young scientists. The head of the laboratory looked through his fingers at these adventures and, moreover, he also stayed long in the rooms of pretty ladies. The only requirement was to be at work at 9 a.m. But it was always possible to get enough sleep after dinner.

And here is the first experimental day. The equipment is tested, the sensor turned on and adjusted. Several people gather in the laboratory room, including the head of the laboratory and one of our powerful sensitives-operators, Leonid, who showed himself in healing and telepathic sessions. An extremely unpleasant feeling: what if everything is in vain? What if nothing works? Maybe the success with Polyakov was by chance? A continuous wavering train of pulses on the screen of the oscillograph. Background mode. The physical sensor can be in this state for many hours without a break, and the signal will be practically constant, with only small variations. I explain the task to Leonid: to influence the sensor which is in the metal box. He asks,

"How shall I influence? What shall I transmit: feelings, orders, or emotions? How do I know?" I reply as a joke,

"Try to come to love it. Imagine that a young pretty princess sits in this cell, and you are her prince-liberator".

Leonid concentrates. His face looks concerned, becomes glassy and absorbed. A minute passes, then another one and five more. A continuing train of pulses pass across the screen, as if nothing happens. I understand that nothing has come out and try not to look at anybody. I again cast a glance at the screen. Time to quit. Why pull the wool over people's eyes... And I suddenly see that the pulses on the oscillograph's screen start growing in size and expanding. Hallucination? But no - a train of pulses is growing actively. The recorder's pen, previously drawing a straight line, also oscillates and definitely turns aside.

"Success", states the head of the laboratory confidently. The silence reigning in the laboratory is broken: everyone starts moving, whispering, and coughing. The recorder's pen goes up and starts making a wave-like curve. Leonid darts a glance at the device and says,

"Well, seems to be enough for the first time!"

It is as if he turns it off: his look becomes normal and kind again, his features smooth, and a smile appears. Oscillating for half a minute more, the recorder's arrow falls down to the initial value.

"Well, congratulations, the first stone is laid. Now we have something to work on", the head of the laboratory stands up and slaps me on the back, "You make the party. Come on, run to the store, and let the girls make some appetizers. This is worth celebrating!"

Victory ! Everyone is happy. Another few months of experiments, and a new stage of work can be organized, new financing for another working year, and the most important - a new interesting phase of research can be started.

After a series of experiments with different operators I switched the device into a 24-hour working mode: the supply voltage was stabilized and controlled by instruments. The recorder was writing the curve of the sensor's signal continuously 24 hours a day. The device was set in a separate room and didn't require any special service. The parameters were controlled and readings were taken two times a day. 5-6 operators were constantly working in the center. Each had 3 hours of work. During that time a sensitive could come into the room with instruments and have some training, looking after the recorder's pen. Or he could work at a distance, marking the time of work. Then the marked moments of influence were compared with the readings of the recorder chart and a conclusion on the presence or absence of effect and its size was made on a ten-point scale.

Such technique made us independent from the researcher's actions - a factor which should always be taken into account when making experiments with people. Nothing was pressing the operators, either: they could work when it was convenient for them without any outer observation or control. I was quite pleased with such operating mode: during the day we could lie in the park with a scientific journal, and process the recorder's curves at night. Night is the most productive time for meditation and scientific studies.

After two weeks of work it became clear that different operators influence the sensor differently. A forty-year-old physician from a suburb of Moscow, Igor, was the leader. Then followed a pretty "witch" (as she called herself) Nelly from St. Petersburg. The effectiveness of influence by others was significantly lower. Once we all gathered at night in the drawing room at a big table. The discussion was combined with tea-drinking.

"Well, now that you all know how capricious this thing in the metal box is, let us discuss how different people work with it", I started, "Igor, you have the highest effectiveness of work, share your experience with us".

"When I influence the sensor, I first of all imagine that this is a part of my organism. Mainly, this is the continuation of my hand. Then I start moving energy, as I usually do working with patients, increasing the energy level in my hands. As soon as I feel that it is under control, I can operate it as I wish. The distance doesn't play any role here. From three meters or from three kilometers. I checked it purposely."

"And how long can you keep this state?"

"The most important is to enter this state. Then it exists. You are already in the flow. It becomes possible to do something else. But not too emotionally. Emotions disturb."

"And what do you feel doing that?"

"Hard to explain. As if swimming in warm water. You feel yourself in a flow. And when you go out of that, the state is very pleasant and relaxed."

"Well, thank you. Nelly, what do you say?"

"I just love it, your sensor. It is like a warm kitten for me - soft and tender. I was trying to find a clue to it for a long time: imagined it a part of myself, read poems to it, and tried to give orders to it. And then I understood that it's just necessary to love it a bit."

"As if it were a man or a child?" asked somebody.

"It is not important. If you love, you love. This is a single feeling. It makes no difference in age, type or character of your beloved. This feeling fills everything around, as ocean water. You totally resolve in it, forget about yourself, and the only important thing is that your beloved feels good."

"Is it really possible to feel something like that to a soulless piece of iron?" asked a girl Nina, with astonishment.

"And why soulless? When Konstantin was doing it, he put so much of his soul into it. So that's why it responds!"

"Interesting, maybe we start mass production of sensors with labels "0.1% of soul," "0.4% of soul" and put the corresponding price!" suggested Igor.

Everyone burst into laughter.

After this discussion all the operators learned "the technology of love." And the results were much more stable. Naturally, different operators had different results. After dinner the effectiveness of work decreased. The same on Mondays. However it might seem questionable. We worked without weekends didn't want to loose the precious time. By the end of summer convincing statistical data on the influence of operators on the gas discharge sensor was gathered 9 . (Fig. 48)

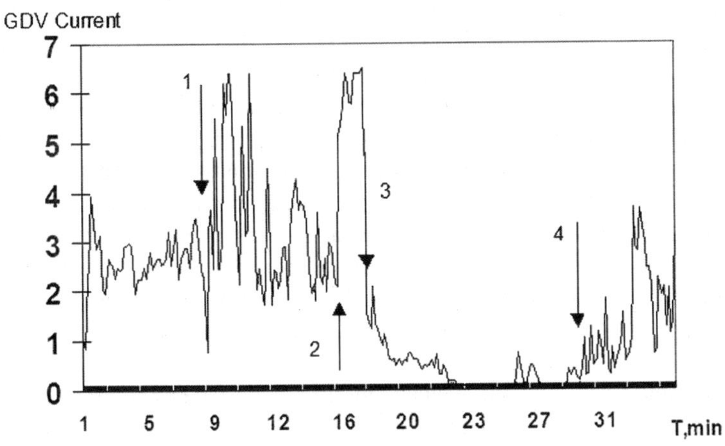

Fig.48. Change of Sensor signal under mental influence.
First 9 minutes - measurement without any influence; moment 1 - beginning of mental influence from the distance, after increase of the signal the influence was stopped; moment 2 - next influence; at the moment 3 the inductor offered to model "the state of death"; then at the moment 4 he was asked "to come back to life".

Later it was found that similar sensor had been tested practically at the same time in the USA by the group of scientists directed by William Tiller 10. That is the synchronicity of the development of events! At that time we didn't even suspect the existence of one another! The history of science has time and again demonstrated that good ideas come to mind with different people simultaneously. Thus it was with the invention of radio, the telephone, the X-ray device, and tomograph. The most important requirement - the society should be ready for these ideas.

At the next stage we wanted to work out selectivity - if the operator was able to influence a particular sensor, placed close to an identical sensor.

We returned to the city by autumn, reported on the results to several commissions, and got the go-ahead to proceed with the experiments. We worked to improve the equipment, checked several other constructions, and tried working with water and biological liquids during the winter frosts.

We had had extensive experience working with microbiological cultures by that time. After a couple of years of work we had managed to substantiate a technique for determining the state of microbiological cultures in the process of cultivation. The technique was based on the application of single voltage pulses and registration of glow in narrow spectral ranges, with the help of photoelectron amplifiers with filters 11.

The main moving force of that work was Galina - a talented ebullient biophysicist. In The Soviet times she successfully defended her Ph.D. thesis on this topic and continued working after that. But then the "perestroika" began, it seemed that scientific work wasn't needed and interesting, the living was very difficult, and Galina left for Germany, to live a calm, quiet life. The technique of investigation of microbiological cultures with the gas discharge method is waiting for its next enthusiast.

Based on our experience we decided to study how the operators can influence biological liquids. Blood and DNA solutions were chosen for the experiments. Some results were obtained. Methods for complex analysis of the state of these liquids began to be prepared: ESR, PH, biological activity, gas discharge glow. Naturally, we had a lot of other work and lines of research in addition to these experiments.

The next summer was spent again in Pribaltica, at an out-of-town lodge. That was very interesting: experiments, the joy of creation when the plans were fulfilled and the next stage was seen. Seminars discussed many questions from the experimental technology to the philosophy of religion. Many interesting people came from the Soviet Union, to share their experience, to work together with us and show their methods. During that time I managed to pass several intensive trainings and master various techniques of controlling bioenergy, both mine and environmental.

On the whole, we established that mental influence on physical systems is possible; both on physical technical systems and on biological systems. In order to do that, operators should be thoroughly selected and trained. For effective work an operator should switch into an altered state of consciousness (ASC), as we call it today. Later we will discuss this question in more detail. Not all can do that, and the effectiveness of work is absolutely different. The same as for musicians: talent and persistent work of many years will show if the person can attain a place in a philharmonic orchestra or a concert career. Professional activity is impossible without painstaking work. Energy-informational work is not an exception. And success results from years of persistent training based on initial talent.

After the analysis of all data obtained during that summer we found that even the effectiveness of work done by the best operators varied from 70 to 80%. During some days it grew to 100% of effect, others - fell abruptly. What did it depend on - who knows? We didn't manage to find some special correlations. It became clear that persistent work should be carried out for many years before practical application. With that conclusion we returned to St. Petersburg and started writing reports.

Apparently, our conclusions were not to the liking of our bosses. Everyone wanted quick effects, triumphal reports, medals of honor on their jackets. When it became clear that the work would take a long time, and the final prospect was rather vague, the continuation of the work was up in the air. What was more, the Time of Changes started: Leonid Brezhnev died, the coffin with his body was carelessly dropped into the grave during his funeral, and the superstitious people interpreted it as a special omen. Then came a short period when Brezhnev's old associates were trying to catch the reins of government one after another; but soon each left, swept off either by illnesses or diseases, and finally a secretary of a regional committee of the communist party, known to nobody, Michael Gorbachev, came to power.

In the course of all these changes, the initial interest in the discovery of new laws of life fully faded, our projects were closed one after another, and all the employees slowly retired. Some changes of the Soviet society were in the air, but nobody in the world could imagine their scale.

"PERESTROYKA"

Much has been written about those years of change. And the historians will further dedicate volumes to the reasons and consequences of the dissolution of the Soviet Empire, and the role of Michael Gorbachev in it. Many people in our country damn him, the others depreciatingly smile. There is a Russian proverb: "Wanted the better, got as bad as usual"... Gorbachev sincerely tried to make reforms and improve the collapsing system without modifying it fundamentally. He sincerely believed in the power of the Communist Party and tried to keep it until the last moment. He sincerely tried to fight against hard drinking, and the local minions were ordered to cut down centuries-old vineyards without hesitation. But all his actions were followed by the consequences, absolutely opposite to his intentions. Having undermined the state monopoly for alcohol, he created conditions for the emergence of a mafia, which then rapidly throve, using the first "vodka" money. "Perestroyka" (reformation) and "Glasnost" (publicity) released maturing hidden forces, ready to fight for national sovereignty, democracy and a multi-party system. The dissolution of the Union occurred suddenly, instantly, and "without blood." And for that monuments should be erected to Gorbachev. Having all the power of a military and police machine, possessing absolute authority, he didn't cling to power, but nobly stepped aside. One can see weakness in that, another can search for intrigues of imperialistic secret services, but from the viewpoint of historical process it was one of the most outstanding deeds of the Supreme Power in the history of Mankind. Are there many examples in history when a dictator has left without fighting, voluntarily giving the power to the forces of progress?

Then the Transition Stage started. Nobody knew what to do, how to do, the old collapsed, the new was being born in the throes. Brigandage, mafia, political and economic shady deals - there was enough of all for 10 transitional years. Were other variants of development possible? For example, in China? Perhaps, yes. It is obvious that Boris Yeltzin's government wasn't the wisest, and governmental reshuffle didn't contribute to the consolidation of the country. Russia fell behind economically, industrially, and scientifically. At the same time, I always recall pages from a sacred book, the Bible.

When Moses led the Jews out from Egypt, he was guiding them in the desert for 40 years before coming to the borders of the Promised Land.

> And the children of Israel did eat manna forty years,
> Until they came to a land inhabited; they did eat manna,
> And they came into the borders of the land of Canaan.
> Exodus.16.35

Why? It is two weeks at the most from the borders of Egypt to the present day Israel, if you walk without haste. Where did they wander with the whole tribe and why? And the reply to this puzzle is given only after many chapters. The God explains to his people through his messenger Moses:

Surely there shall not one of these men
of this evil generation see the good land,
which I sware to give into your fathers.
Deut.1.35

Moreover your little ones, which ye said should be a prey,
and your children, which in that day had no knowledge between good and evil,
they shall go into thither, and into them will I give it, and they shall possess it.
Deut.1.39

Moses was deliberately guiding his people in the desert. He was waiting until the new generation would grow up. Moses spent forty years to bring up a new person, a new generation, born free and never feeling themselves slaves.

The problem of transition to a new system wasn't mainly economical, but, first of all, spiritual. The Soviet system was striving to turn people into obedient machines and that turned out well in many respects. The people lost their ability to think, independently make decisions, and be responsible for their actions. They forgot how to work persistently. Time is needed in order to bring up all these qualities, to create a Russian citizen. And comparing the situation of today in the Twenty-first century with the previous situation, we can unambiguously certify that the progress is present. Slowly, with pain and groans, Russia rises for an active life. People are learning to do business, overcome the barriers of officials, independently distribute resources and rely only upon themselves. Of course, all that isn't easy, especially in the absence of a normal banking system, business infrastructure, and state support, and with creeping inflation and unrestricted bureaucratic despotism. At the same time, store shelves are crammed with goods and the stores don't suffer from an absence of customers. Russian speech is heard in any part of the world, and Russian speaking personnel are specially hired in many resorts. And the most important - young people do not aim at going abroad by any means, as it was before. Many prefer living a more complicated life, but in their own country. Russia takes its place among the leading countries of the world with increasing confidence.

MY ODYSSEY

To me, changes taking place in the country created optimal conditions for realizing ideas which I had matured over a long time. In the Soviet years a scientist had no right to take any step without the official approval of the directors. The signatures of members of a special committee should be obtained in order to send an article into a scientific journal. That committee evaluated if an article was suitable for the open press. Any unauthorized initiative was seriously punished. Not to mention that, as a scientist, I was prohibited from going to capitalist countries.

After leaving the closed institute, I tried a whole series of occupations. First, I tried to start professional alpinism. This took place by chance, as everything else in my life. I'm staying at home after dismissal and thinking: what do I start? It was quite difficult to get established in a scientific job anywhere then, and practically impossible to continue investigating the Kirlian effect. But I had distaste for everything else. There were no serious problems with finance then: I earned money giving lectures in the society "Znanie," on various topics of natural science, from space flight to the philosophy of religion. In spring time it was always possible to earn money elsewhere by steeplejacking - this was the main source of income for many of my alpinist friends. But I longed for doing something more substantial.

And here my alpinist friend calls and, among other things, asks,

"And what do you mainly do now? "

"Nothing special in fact. I've quitted my job and haven't entered anything else."

"Perfect! Come to us, to the department of alpinism. You can give lectures and guide groups to the mountains every three months."

The idea was alluring, and in a week I was giving lectures on rescue operations in the mountains. Indeed, everything appeared to be as described by my friend: easily accomplished lectures, regular hikes to the beautiful mountain regions, and winter ski trips. But after half a year I realized that I got far less pleasure guiding groups of young sportsmen, training them to work in the rocks and on ice, than in making sport climbs. I felt myself at work in the mountains. And that prevented me from enjoying the mountains. But then - another turn of the spiral. Out of the blue, I was offered the position of acting director of the institute where I had worked at the department. A concourse of a whole series of circumstances led to that offer, and my candidacy turned out to be the most convenient in that moment for the ministry. I was occupying that post for less than a year, but had time to see enough of the savage customs reigning in the ministry. It seemed that the main pursuit of the officials was weaving plots, each worthy of Alexander Dumas's pen. That was a perfect schooling which disinclined me forever from doing an official career. Besides, I was more and I missed more and more the world of technologies, electronic "pieces of iron," intellectual games, and the living, always changing Kirlian images. Therefore, I left the warm place at the institute at the first opportunity and came back to my pieces of iron.

Then the vague years of Perestroyka set in, the time when there was a lack of simple products in the stores and the near future was absolutely uncertain. Several times the country was on the verge of a catastrophe. Many prophets foretold a civil war and even gave the date of its beginning. Such prophecy failed. We survived. Angels protected us. Russia wasn't doomed to get into the series of civil war again in the Twentieth century. Now the historical attractor was different.

But shots were heard in the streets at night. Outbursts of brigandage. I had to carry a loaded gun in a pocket. Although, thanks God, I didn't use it. I was thinking of leaving to live abroad. But, apart from all that, I had my Business. Business, which gave an opportunity to start the research again with full enthusiasm.

I started my first company in 1989, as soon as the first law on small business was issued. I hired a few employees - former colleagues at work, opened a bank account and began activity without investing any money. We started taking orders for the development of electronics and production of equipment - this was interesting, but gave practically zero profit. Then we organized the production of originally designed crystal lamps - and this went on more successfully. I tried to establish relations with Finland, Sweden - spent a lot of time, gained certain life experience, but no income.

The following anecdote gives an idea about the business of that period.

Two businessmen meet in the street. One asks another one,

"Do you need sugar at 2000 per ton? "

"How much?"

"A wagon."

"When?"

"On Friday."

"Can I take two wagons?"

"OK, only for you."

"Deal. Give the money when see the goods."

"Bargain!"

And the businessmen take off in different directions. One - to search for money, the other - to search for sugar.

Something went on successful, and I started spending the very first income for the creation of a new Kirlian instrument. Previous years of scientific experience enabled me to formulate the parameters to be put into the construction of this instrument, but over many years life circumstances had prevented me from realizing these ideas. As soon as an opportunity appeared, I took the dusty folder with calculations from a shelf and started developing the new generator. What for? It wasn't clear then. No profit could have been received from that. But I had an inner feeling that this had to be done.

The first samples were quite operable, but didn't provide the required parameters. The fact was that the glow could be obtained by means of any simple high-voltage generator - even by a kitchen piezo-lighter. Everything would be shining and sparkling, at that. Many experimenters had chosen that method during two centuries. But simple generators gave very unstable voltage, and no reproducibility. Everything was glowing, but no quantitative information could be derived from that glow. Stability at the level of 0.1% variation of all parameters was needed. And, secondly: our physical investigations proved that not a drop-down sinusoid, but short single pulses were the optimal stimulus for glow. But it was not easy to reach these parameters.

Models not satisfying strict criteria were left aside, one after another. I thanked a group of developers, calculated the resources, and started searching for the next group. As it appeared, no one was developing the needed high-voltage equipment. Neither powerful generators, nor sources of direct voltage for photomultipliers. Once I was asked to apply to some developers of electro-shockers. At that time - beginning of 1990's, the whole Russia was arming, and electro-shockers were among the pretenders to legal weapons. That was a device giving a few kilovolts with low current. Impossible to kill anybody, but quite painful. Then shockers were finally found to be dangerous. Moreover, these devices were used by bandits and not by peaceful citizens, and so they were prohibited. But at that time there was a company in St. Petersburg developing and producing these devices. I contacted some engineers from that company, and after long discussions, trials and errors we managed to create an instrument with quite satisfying parameters. With this instrument the experiments could be started again.

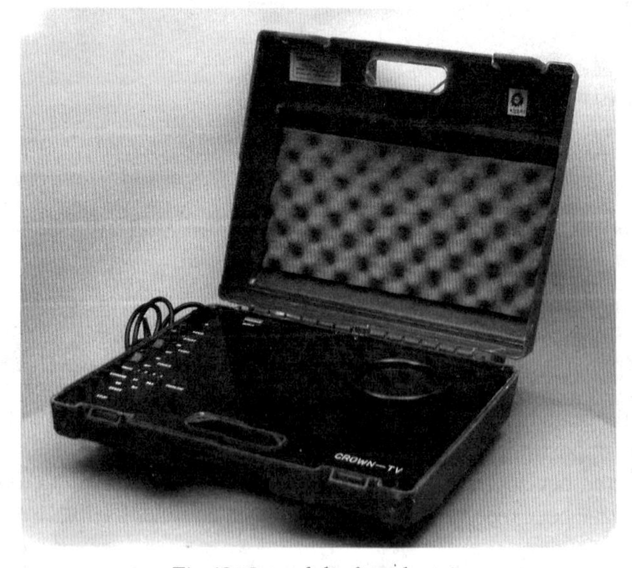

Fig.49. One of the first devices

NEW STAGE

We didn't have to start from scratch. We already had previous experience, relations and connections. Having made the first few generators, I called couple of my friends who was medical doctors and we started investigating patients in several clinics. The diagnostics were based on Peter Mandel's tables, which we translated from his book. Selection of data and accumulation of experience started. Folders began to be filled with photographs.

At the same time I tried to start developing software for processing Kirlian images. There were practically no PCs in Russia that time, so I found a group of programmers in one of the higher institutions. They had experience in image processing, and agreed to start developing software for reasonable money. The first idea was to calculate the areas of Kirlian photographs based on Mandel's sectors.

We spent half of the income obtained from other activity and almost all free time on this work. We couldn't even think about any support from the state then. Higher learning institutions were in a difficult situation: financing was not enough even to pay for electricity and heat supply. In winter the classes were often held in unheated rooms, and students were sitting in coats and hats, warming up the freezing ball-pens in their hands. A professor's salary was lower than the cost of living. Research institutes were emptying, the employees were searching for an opportunity to earn money for living, and free rooms were rented out in order to receive at least some means.

However, creative power was boiling over. Many scientists who were no longer bound with official agreements, reports, and salary were investigating questions which were interesting to them. From time to time they gathered at conferences and seminars and discussed the ideas of energy, information, eniology, torsion and scalar fields. Dozens of people were gathering at these meetings, discussing new scientific directions for hours.

These meetings were an echo of general changes taking place in the society. The social order changed totally after a few years, all spheres of life changed, and people stopped being afraid of the new. A new life style opened the way for the perception of a new paradigm of science. The atmosphere of conservatism, typical of totalitarian states, was changed for the era of search and an open attitude to ideas which had previously seemed to be seditious. The air of freedom was fascinating and gave birth to new notions. The collective consciousness of the society survived serious transformation, and from being stark and frozen it turned into flexible, dynamic and susceptible to everything new.

The activity of two people, Alan Chumak and Anatoly Kashpirovsky, greatly contributed to this new open consciousness research. In the beginning of the 1990's these people were allowed to perform on national television with sessions of mass influence. Their form and method of influence on the masses was absolutely different: Kashpirovsky, ascetic, and rigorous by appearance, played a black magician from Oriental fairy tales; and the good-natured, roundish Chumak had the habits of Bulgakov's cunning cat Begemot. Kashpirovsky was inculcating his program upon people, and that was effective, although not always harmless; Chumak was more silent, spreading his kind energy to the audience. Large public opinion polls demonstrated that in both cases there were a lot of positive effects: migraine headaches passed, the state of people with chronic diseases improved, and disturbed sleep was restored. There were thousands of such responses. Kashpirovsky and Chumak started performing at large stadiums, and the tickets were sold out. Others followed their lead. The situations were comical.

Performances of "Baba Njura - a sorceress in the fifth generation" or "Lama of the third consecration Sidorov" started in many large cities. "Treatment of all diseases. Removal of evil eye. Admission free." I couldn't understand how these "babas njuras" made money - the entrance was free. But then I was told about that.

People came out of curiosity. Why not if the admission was free. "Baba Njura" or "lama Sidorov" were mainly telling about their success in healing most terrible diseases for an hour and a half, and sometimes they even demonstrated "patients" jumping with joy. In order to be treated successfully one was to regularly consume water charged by the magician or wear amulets impregnated with his positive energy. By the end of the performance those who wished could purchase the magic objects in the foyer. And they did, since the price was low. What's the difference: 20, 50, 100 rubles for a healing subject. And what if it helps? And people eagerly bought up the goods. Thus, bottles with water at the cost of 1-2 rubles were sold for 20, providing swindlers with 900% profit at minimal investment. One session brought thousands of dollars, and no taxes were paid.

Soviet people were taught to believe the press. The main newspaper "Pravda" (Truth) was issued in the Soviet times, and all the materials published there were taken in as the truth of the ultimate authority. If "Pravda" was praising somebody, the person was awarded with an order, if criticized - the person's career and sometimes freedom came to an end. The "Pravda's" opinion on any issue was an official viewpoint of the Central Committee of the CPSU and was taken as a directive to action, whether it concerned industry, science, or culture. The government was clearly determining what was good and what was bad. The bad was simply destroyed: bad scientists, bad painters, and bad poets. And the people truly believed that everything should be like that.

Therefore, when "Baba Njura" was pathetically telling how she was using the ancient force of her great-grandmother from the times of Tzar Ivan and healing all diseases, people believed. What was more, nobody asked money for that. One could voluntarily pay or not pay. Mainly they paid.

Later on such mass sessions were prohibited, but they had done their part. The example of a large number of people demonstrated that the word can influence the state of health. Naturally, a question arose: wasn't all that a placebo effect, i.e. the effect of self-suggestion? Suddenly this question passed from the academic circle to the practical plane.

LIGHT AFTER LIFE

One of the topics stirring my imagination for many years has been the question of the life of the soul after death. In the Soviet times we were far from being religious. Going to church was not considered good and could sometimes lead to fatal consequences. The believers could not make a serious career, paths to diplomatic work or management were closed to them. Searching in libraries I mastered a course of philosophy of religion, from Schopenhauer to Frazer, studied the Koran, Bhagavadgita, and Cabbala in detail, let alone the Bible, and gave lectures on all this in the society "Znanie." At the same time, this was a purely academical interest and there was no special spiritual difference for me between the sayings of the Prophets or koans of Zen.

Books by Raymond Moody, Elizabeth Kubler-Ross, and Stanislav Grof awakened great interest in me for the questions of transformation of the soul after death. But all the evidences were based only on the stories of people who went through an apparent death. There were no experimental data. And natural questions emerged, "If we can change the energy-informational state of a human during his/her life, then what would be after death? How would the Kirlian images look and how would they change with time?"

I started discussing these topics with my friends-physicians and came to a conclusion that it was quite possible to organize measurements of the Kirlian glow after death. But in order to make such observations it was necessary to gather a certain amount of data, and that required a lot of time, a whole group of specialists, equipment, computer processing, and so on. And all that was not only time, but also money. When we calculated the very minimum it was quite a significant amount. We scratched our heads all together, discussed who might give money for such a project, and left it off as an interesting idea.

I had heaps of projects from different spheres of activity during those days. Some generated official proposals and started their journey in the ministries and departments, where they disappeared without a trace. But with time some ideas were realized and gave interesting results. And now the number of ideas significantly exceeds real possibilities to fulfill them. But at the same time, with a glass of red wine, I regularly tell my friends new ideas.

Thus it was with the idea for after-death measurements. So, once I was sitting with a friend from Finland, Helge Savolainen. This is a very interesting and talented person. He had worked in the Ministries of Finland for many years, and then he successfully started his private business. Having met a couple of talented healers in Russia, he got interested in their work and presented them in Finland. Again complete success. Helge also has another hobby. He photographs women. Any women. Fat, slim, or short. And he makes beauties of them. Mermaids, fairies, and druids. He opens their inner essence and makes them believe in their own beauty, takes off depression, makes them forget disappointments and find new life impulses. His portraits come alive and one can fall in love with them (But, as a rule, this is a bitter experience. Be careful with Galateas! The illusion of art is much sweeter than real life!)

Together with his wife Victoria - a talented Russian healer, Helge often lives in St. Petersburg. We used to sit with him in the kitchen, drinking beer and discussing different projects. So, once during such a leisurely evening Helge said,

"Konstantin, I've made a successful business deal, and want to do something good. How much money do you need to perform experiments with the after-death glow?"

A month later we started our first measurements.

I will not describe the details of this work here. In 1996 my book "Light after Life" 13 was published, where both experimental data and various ideas about that were given. The results became popular in the world. The book was republished in the USA, articles were issued in many countries, and TV programs were made. The main conclusion of the research was that energy-informational activity of the body after death doesn't stop for several days, i.e. some informational structure is shown by the body. An important experimental fact was found: the character of activity in this structure depended on the type of death. In a case of calm, natural death the activity continued for 2-3 days; in a case of momentary death from a traffic accident a powerful outburst of energy, a fountain of energy was observed, lasting for a few hours only; and in a case of suicide, murder, or smothering, oscillations of activity continued for 5-6 days and sometimes didn't fade away during the whole period of measurement. (Fig. 50, 51).

I had to transport the diseased or injured alpinists and help doctors during postmortem examination several times during rescue operations in the mountains. And that caused deep emotions every time. Even if the distressed person was an absolute stranger. The experimental data obtained completely corresponded to religious and esoteric notions on the transformation of the soul after death. We discussed this topic with the priests many times. All of them took kindly to our work and emphasized its importance. We were discussing this topic with father Sergiy in a TV program broadcast by the central Russian channel on the Easter eve. The in-depth study of this topic, reading of many books, including those written by the fathers of the Orthodox Church, have been the basis of my personal sincere attitude to this belief.

At the same time, this work posed much more questions than answers. We observed a strong influence of the dead on the energy state and health of the experimenters, a negative influence. We observed the connection between the world of the dead spirits and the world of the alive - both on the basis of sensations and indirect measurements. And the main point - we made certain that such "subtle matters" could be investigated by means of modern science.

In recent years work on brain activity mapping during ecstatic sensations such as prayer or meditation have attracted much attention. It has been found that certain brain areas are excited and a person can come into special states: altered states. Recall that I've already mentioned that topic. We will yet come back to it. So, what

1.

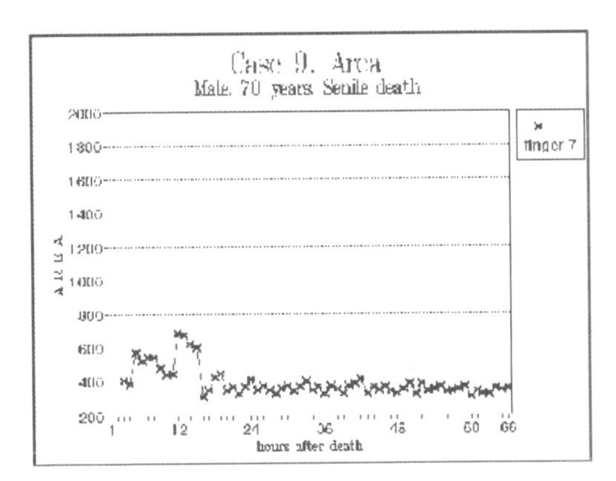

2.

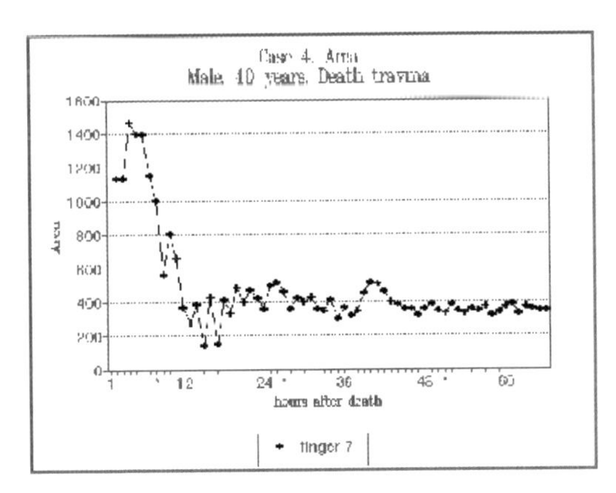

3.

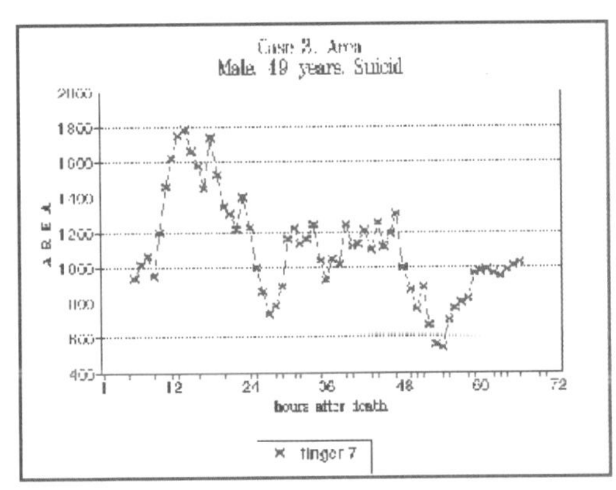

Fig.50. Time dependence of after-death energy.

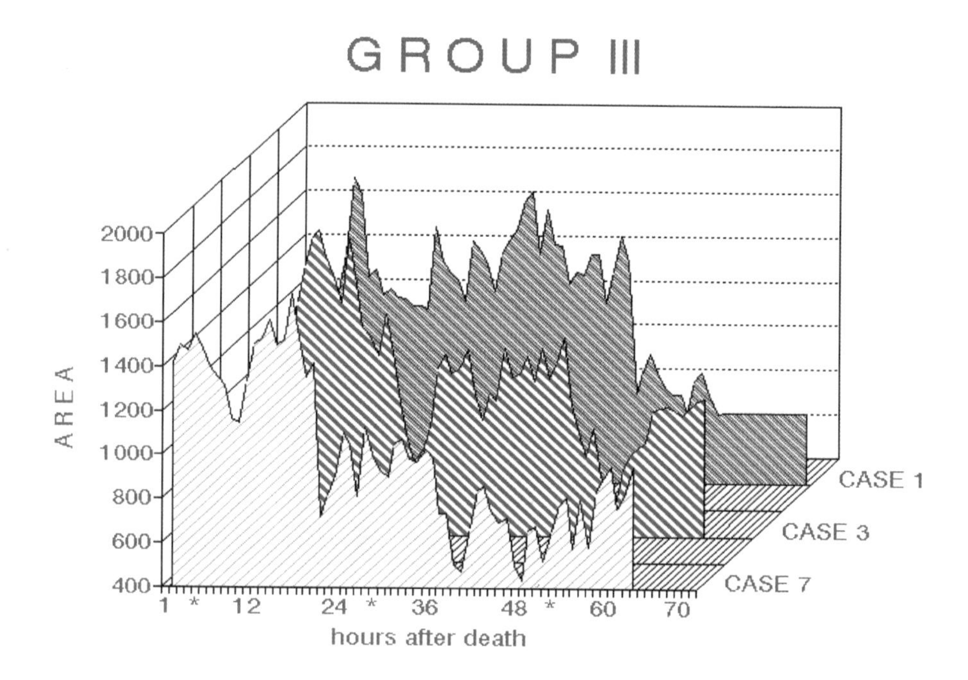

Fig.51. Time dependence of after-death energy.

is primary: excitation of separate areas of the brain, causing a sensation of contact with the Supernatural, or information, sent from the Highest Mind, from the Creator, and realized through our brain? Who were Moses, Buddha, Zaratustra, Christ, Joanna of Arc - psychically unstable people, inclined to the development of brain impulse activity, causing visions, which could hardly be distinguished from reality, or the Messiahs of the God's will? Who are genius poets, writers, composers - diseased people living in the world of twilight imagination, or sensitive indicators of the epoch, feeling Collective Informational flows with every body cell and capable of expressing them in their creative activity? Many measurements demonstrated that spreading and local epileptiform activity is developed in the brain during healing activity. But this is not a clinical diagnosis. This is just a similar character of brain processes. The "Yurodivye" ("God's fools") were considered to be messengers of God since ancient times in Russia and not for nothing. Even the tsars listened to their words. The great mystifier Salvador Dali once told: "The only difference between me and a mad man is that I'm not mad". So, what is our brain in fact: a source of information or receiver-transmitter, encoding and decoding signals from outer space, including those from the space of Higher Spheres? What is Church - one of the institutes of mankind invented for the restraint of masses, or the Spiritual Messiah, bringing the Word of God to people and strengthening the spiritual unity of a nation?

Now we can't answer these questions definitely. Being a believer, I strive for finding confirmation of primacy of the Spiritual over material. As a scientist I try to do that objectively, based on experimental facts. I'm sure that first of all convincing evidence should be obtained, and only then we can build theoretical notions. Otherwise it's just an idle talk which has prevailed recently and has never yet brought any special use. There have always been chatterers, but what has remained from them except for empty beer bottles?

And another observation regarding our experiments with the after-death glow. This topic is very delicate, requiring most accurate attitude both when the experiments are carried out and discussed. After my lectures I often heard the question, "Weren't you afraid to start such a topic?" And again a situation from the alpinist practice comes to my memory.

FEAR

It happened in the mountains of the Pamirs during my third summer season. We had a training climb, and after that together with a friend we decided to start a difficult wall rise. A girl from our section stuck on us to go, too, and after a short preparation the three of us left the camp.

A rocky way went up to the pass, from where the view opened to a green valley with turquoise spots of lakes. We ascended along stony valley, in the end of which the dried brook-bed led to the vertical rocky wall - the aim of our trip. The Pamirs' sun was hanging high up in the cloudless sky, and the warm air quivered and seemed to be dense to the touch. We reached the brook, and found with satisfaction that the streams of icy water were streaming down the rocky shoulders.

Each hour we went higher and higher and a slight wind appeared at 4000 meters which was of immense help. We had got out to the platform at the foot of the wall by night, threw off rucksacks, and started viewing the route of the forthcoming climb. Everything was absolutely clear and corresponded to the description: a wide vertical cleft convenient for climbing went up right from the place of our bivouac. After late dinner we got into sleeping-bags and, having exchanged anecdotes and jokes, got ready for sleep.

I was lying, my eyes closed, but sleep was not coming. Fear was moving in the stomach as a disgusting worm. Namely, anxiety. I was physically feeling that we were lying under a huge vertical wall with hanging ice on the top; we, tiny pygmies, who dared clambering that wall. What was there ahead - steep plain rocks without any little places to grasp on; stones flying from above with the speed of a shot; a kilometer abyss under foot and thin webs of ropes, fastened to the unsafe hooks driven into the rock. Catastrophes in the mountains, which we regularly discussed at the seminars on safety, were recurring in my memory. Finally, I fell asleep, but the sleep was restless and short. In the morning we drank tea, prepared our rucksacks, tied ourselves with ropes, and started climbing.

During a difficult rise one alpinist goes ahead, carries a rope after him and drives in hooks in order to provide his safety: in case of falling down he flies to the first hook and hangs on the rope held by the partner below. Naturally, he hangs if the hook stands the jerk, if the partner carefully secures the rope, and if the rope doesn't break from the load. During the years we had a lot of different situations. So, it is better if the first alpinist doesn't fall down. Having passed a rope (40 meters - height of a ten-storey house), he drives in the hooks, secures the rope and meets the rest of the team.

The first ropes were climbed quite easily. The cleft was steep, but convenient for climbing. The sky was clear, and a wind was light. In a couple of hours we got onto a high shelf, from where a marble wall went up with little places to hold on and cracks convenient for driving in the hooks. We worked one after another with Tolya: his rope, my rope, the girl was securing from below. Every meter upward was careful and precise: we were literally touching with eyes and fingers every little place to grasp on, every ledge, and every crevice in the rock. Maximum concentration and attention, no time for emotions, feelings, and "sightseeing."

By about 6 p.m. we had passed the marble wall and came close to what appeared to be a huge fireplace, forming something like a cave with a wide flat platform. A perfect place to put up a tent.

In a few hours we were already lying in the sleeping-bags, having drunk tea with dried bread crusts and sausage. Each of us was tied to the main rope put through the tent and fastened to the hooks driven in the rock. Even at night we had to remember that we were lying on a tiny platform in the middle of the vertical wall.

The wailing of the wind made us wake up at night. The evening radio contact had transmitted that the weather would go bad in the coming days. It was our bad luck that the weather forecast came true earlier than it had been announced. We were lying half-asleep the rest of the night, listening to the gusts, the shaking and moving of our tiny house, and the rustling of the falling snow.

When the dawn had broken we looked outside and saw that all the rocks were covered with snow pellets. The sky was lowering, the ragged clouds were flying in it, and our hands were cold even in the tent. We weren't in the best spirits. The whole middle part of the wall with a difficult vertical rise was still left. One thing is to climb a dry warm rock, and it is something different to climb cold icy ledges. That's worse than driving a car on worn-out summer rubber tires along an ice-crusted road. But we had no choice: the descent was very difficult, and how could we come back to the camp having given up to weather? Nobody would have said a word, but that would have thrown us to the back ranks in a non-official rating. Tolik put on the equipment and moved up. I and Lena were securing the rope.

The ascent went on very slowly. A part which could have been passed within an hour in good weather took three hours. The rocks were frozen, we had to shake off the snow in order to find a little place to hold on. The ropes quickly got wet, and the hands were badly freezing. What was more, a biting cold wind was blowing. In the middle of the day the three of us gathered at a small ledge in the rock. Dark bastions were beetling over. It was obvious that we couldn't reach the ridge that day.

"Well, we got into trouble", said Tolik, "and it seems we'll have to spend the night sitting here. Would be good to find a place to hook a tent. Well, do take the equipment and work up. Now every hour is precious."

He handed me a hammer and hooks. And then I felt a deep inner horror. I had been mountaineering for three years by that time and had reached a good level. I had gotten to a feeling that there had been no impassable routes for me. That had given me an air of my own strength and had led to losing inner control. I had become madly bold and had climbed difficult vertical rocks without anyone securing the rope. Euphoria from my own mastery, pride - one of the most serious sins in Christianity. And once in early spring my hand had slipped off from a little place to grasp on, I had overbalanced and had fallen on my back from the height of 15 meters at training rocks near St. Petersburg.

What had saved me then - I don't know. Having made sure that there had been no fracture and having rested under the shadow, I had slowly gotten home with the help of my friends. And at night I had woken up with a feeling that the end had come. Pain had been everywhere, it had been beyond the limits of normal senses, and in order to stop it any means had seemed to be good. Even the final one. I had been X-rayed and injected at the hospital and had come back to this world by morning. Apparently, some internal hemorrhage had opened, as the doctor had concluded. In a month I had been absolutely healthy and had come back to training, but the feeling of a horror of death had remained deep inside me. And so it rose again, causing almost sickness.

"I can't, Tolik, sorry. I feel very bad. I feel very bad."

Tolya peered at me and understood everything. He had served in airborne troops; he had spent two winters in Antarctica, and had seen a lot in his life. Without a word, he put on the equipment and started climbing up.

We spent two more days on that mountain. The weather didn't get better. Lena lost courage completely and we had to literally pull her on the rope. We spent the next night at a wee ledge, sitting, snuggling up together for warmth and holding the working primus stove in our hands. Only by night of the next day had Tolya climbed the next rope and cry out from above: "On the ridge!" That meant that we had passed the vertical wall, then we had to spend several hours to climb to the top and a long day to go down to the glacier.

Neither Tolya, nor Lena said anything in the camp about what had happened. Our rating went up significantly: not many teams were able to complete a wall ascent in such weather. And I understood that only Tolya was the one to be obliged to. He was almost not talking to me.

I decided that I had to overcome my fear. But how? After some consideration I asked the head of the camp to let me climb a few easier routes together with... Lena. As if we had had to collect climbs for a sport class. In fact, I realized that I would have to rely only on my own forces.

And that was a real trial. We started from relatively simple routes, little by little making them more difficult. At first every step up troubled us - it was frightful... But I was making myself go up, overcoming sickness. And at some moment the fear disappeared. The brain realized that caution combined with experience enables us to overcome rocky bastions and safely come down. This was a turning-point in my mountain training, and Nature, as if it felt that, started favoring me. Of course, there were injuries in the mountains, I happened to get into thunderstorms and snow slides, but I came back safely every time and was impatient for the next adventure.

Fear is an essential element of life. This is a protective mechanism of our brain guarding against reckless behavior. But in many cases this is a hindrance blocking the actions of a person frozen with fear.

It is a well-known fact that 90% of drowning people are lost because of fear: in an unforeseen situation a person gets frightened, panics, looses forces, and - drowns. Once local citizens at Ceylon showed me a column of rippling surges, coming above the water.

"What's that?" I asked.

"This is an underwater whirlpool. From time to time they are formed near the reefs at ebbs and flows. If a human gets into such a weel, it swallows him up. Many Europeans have drowned that way."

"So, it's impossible to escape from this whirlpool?"

"No, why not. It's necessary to relax and let the whirlpool carry you down, to the bottom. There the swirl ends and you can swim out of it. I have happened to experience such a situation several times, and, nothing", he smiled broadly, as if inviting being glad together with him.

Such are the majority of situations. There were cases when homicidal maniacs were weak undersized people, much weaker than their victims. But first they scared somebody, and then did everything they wanted with the paralyzed victim. Recall the python, captivating a mouse or a monkey by its stare before gripping it in its twines.

In order to overcome fear it's necessary to imagine what threatens you and try to estimate the reality of danger. And then calmly think how to avoid it. You can start to resist, fight or call for help. The main thing is to overcome the first impulse of freezing fear and to rationally look at the situation. In most cases it works well.

For the experiments with the dead, I was quite rational in organizing them. There are a great many people who constantly deal with the dead: priests, pathologists, and grave diggers. And all of them have continued doing their business for many years. The question is not just in what you are dealing with, but also what your attitude to it is. Say, in thrillers developers of bacterial weapons appear as awful monsters ready to destroy Mankind. Actually, these are nice people, professionals, doing their business and fulfilling their patriotic duties.

Stop, and what about hangmen and professional killers? Does it mean that if they serve at public service, they are worthy men, and if not - bandits?

Officially, this is true. But we approach the sphere of other questions: moral values. There is a system of commandments, and it is fully given in the Bible for Christians and in the Koran for Moslems. If a person follows it in his inner estimation of actions, he can sleep well and his conscience will be clear. Without this system of values the society starts turning into a crowd of bandits and thieves, ready to be at each other's throats at the first opportunity. My Finnish friend Lassi Lehtomaki enjoys telling the following parable,

"One person got into the Underworld. There he saw a big hall full of angry people dying of hunger. A large cauldron full of food was standing in the middle. A big spoon with a 1 meter long handle was tied to each person's hand, so that a person could get the food from the cauldron, but couldn't put it into his mouth. But as soon as the food got onto the ground it disappeared. And the people were wandering in the hall, hungry, angry, and ever dying from starvation.

Then the person got into Paradise. And he saw many people in a big hall; they were happy, glad, and joyous. A big cauldron full of food was standing in the middle of the hall. A big spoon with a 1 meter long handle was tied to each person's hand. And people danced and were happy with one another.

So, what is the difference?

In the Paradise people fed each other."

All actions committed in accordance with moral commandments give good results. Although, perhaps, this isn't a direct process like "given - taken." It is much more comfortable to live keeping in mind the rule, **"Every good doing will be punished".**

This paradox has a deep meaning. If we commit a kind doing and look forward to a requital, in our heart of hearts we don't think about the good which we give, but about the reward which should follow. And this is greed. The good should be absolute and without return. If you gave money to your friend, forget about it. If you can't forget - don't give. If you are saving a fox-cub from a trap don't expect that he will be licking your hand with gratitude - he will run away at the best, bite you - at the worst. It is only in soap operas that grateful orphans throw themselves on the neck of the hero who saved them from drowning in the pond. In real life the saved person starts swearing because the "hero" didn't also save his hat out of the pond. Expect ingratitude when you help people. This is typical. The contrary is an exception.

So, does that mean that one shouldn't make good deeds?

On the contrary. The more the person gives, the more he gets. The more good deeds he commits, the more success he has. But not directly. And only if his deeds are sincere and unselfish. Then he is successful, but absolutely in something else.

Therefore, the life criterion is not the small chatter of everyday chatter, vanity and the kaleidoscope of phenomena, but global result over a long period of time. God is long-patient and forbearing, but finally does to each according to his merits. However, there are also effects of individual and collective fortune, epigenetic heritage, influence of the man-caused and social environment, but these are other questions.

Our experiments were done with clear hands and good conscience, with deep respect to the dead and the traditions connected with them, without any selfish ends. And the best criterion is that many wonderful results have been obtained within recent years, and that a kind hand, helping and protecting, has accompanied all our activity.

TWIST OF SPIRAL

More and more physicians were participating in our Kirlian research. I regularly gave lectures at scientific meetings and conferences in different cities of Russia, after which physicians came to our Center, studied the results, and then some of them started using Kirlian photography in their practice. We began producing generators, and after some time this activity finally broke even. But it didn't make any profit then. It was rather the other way round. There was business activity creating income, and scientific interests consuming that income.

A stark segregation of the society started in Russia. Successful businessmen, mostly from the former party elite, were trading the "Mercedes" cars one after another, while most of the people were wondering what they would eat the next day, and if that next day would come, at all. Therefore, our small flat where I lived with my wife, and later with our little daughter, was a meager home.

I remember that once colleagues from England came to St. Petersburg. They found me by my articles and wanted to get acquainted with me. They had just a few free hours late at night. So, we invited them to our home. And why feel shy! We live as we live... We drank tea, talked, after which the English businessman asked: "And where is your study?" I did not have to stand up, I just moved aside the curtain and showed the place behind the wardrobe, where just a computer on a suspended table and a chair could fit. At first they didn't believe, but then we started laughing all together and they asked to take pictures of me "in the study."

But conditions of life didn't prevent us from enjoying ourselves, rushing around the city in the old ramshackle "Moskvich" car, and going to the Crimea to the rocks in summer. The time was good and joyful.

Especially pleasing was that gradually more and more data on the diagnostic importance of Kirlian photography was gathered. We checked Mandel's tables, having changed them slightly, but refrained from taking images of toes. Mandel's interpretation of energy cycles, which he extracted from simultaneously taking images of fingers and toes, was not clearly understandable from his book. I wrote him a letter, got a reply with an invitation to a workshop in Germany, but we couldn't afford such a trip then.

Later on I met Peter Mandel and made friends with him. He turned out to be a very sweet person, a visionary, and an inventor. A real creator. His system "esogetics," produces wonderful results in his hands. This system is based on a mixture of philosophy and mysticism, and describes circulation of vital energy in the organism. The system is very original and extremely difficult to understand. Mandel regularly conducts seminars, and there is a group of people in the world who use his system with enthusiasm. And they effectively apply Kirlian photography using Mandel's system of interpretation to analyze patients. At the same time, having no Ph.D., Mandel doesn't show any interest in scientific approbation of his system. He lives and works as one of the last free artists in the world of complementary medicine.

The interest in the Kirlian effect was growing. An idea came to me to write a book on that topic, summarizing all data obtained by that time. And as it usually happens with everyone who just starts working with a PC, as soon as I wrote about one third of the book the system crashed and all the work was lost. We mourned, drank port after that, and I began everything again from scratch, saving each article on floppy disks and paper.

In half a year of persistent work the manuscript was ready. The only thing we needed was to find money for its publication. Publication of scientific literature in Russia completely stopped at that time, and it was possible to issue a book for cash only. I tried to apply to banks, but they only smiled at me: at that time everyone was busy with building financial pyramids and pumping the money over to the West, no one was interested in scientific activity.

Again Providence intervened. Once I visited a friend, a businessman who used to be a good scientist in the old days. We were sitting, drinking cognac, and between this and that I told about my problems with the book. He asked,

"How much do you need?"

"About one thousand and a half. To make galley-proofs and issue, at least, 1000 copies." He went out to another room and came back in 5 minutes, bringing a wad in his hand.

"Take it. You can return it later. Hope your book will be profitable. And don't forget a copy with the author's signature."

In a couple of months I was already holding in my hands a book called "The Kirlian effect" smelling of printer's ink. It was much in demand and already in two years became a bibliographical rarity. I have only two copies myself.

Once Gennady N. Dulnev, professor of St. Petersburg University SPITMO (St. Petersburg State University of Informational Technologies, Mechanics and Optics), attended one of my lectures. He approached me after the lecture and invited me to visit the university. After a couple of meetings and hours of conversation Prof. Dulnev invited our scientific group to his laboratory.

This was a prestigious offer. There is a big difference between a small center developing some approach and Technical University, one of the most prestigious higher schools in St. Petersburg.

Prof. Dulnev is a person with a very interesting pattern of life and original mind. He used to be a mountaineer in his youth and reached the highest grades in the Soviet sport of 1930's. Having started active scientific and educational work, he was appointed rector of the present Technical University and guided the work of the institute for many years. During that time the institute became widely known not only as the main place for training opticians in the Soviet Union, but also as an outstanding scientific center. Many space optical systems, the first lasers, and holographic systems were created in ITMO. Prof. Dulnev was also the head of the department of thermal physics and wrote a number of monographs in his specialty.

Then he chanced to meet Ninel Kulagina. This woman possessed an amazing gift to control some unknown forces. Putting her hand over a person's skin and smiling she could cause a second-degree burn; light objects would move from place to place under her gaze, and she could influence a radiculitis, rash or headache. Many saw her skills, but nothing resulted from these encounters, except for cries of astonishment and suspicion in some charlatanry. Official sciences claimed that those things couldn't be because it could never be that way, so nobody was really interested in that. Prof. Dulnev was a rare exception. He not only got interested and spent many hours with Ninel Kulagina, but organized scientific experiments with her participation! That was a great risk for those times. Party bosses were calling the rector of ITMO Prof. Dulnev and asking,

"Gena, you are said to have started some devilry there? What's up with you? Not enough work? Watch until it brings you somewhere..."

A usual official for those days would have stopped all the experiments and burned up all papers after such a warning, but Prof. Dulnev was a real alpinist and had always realized his projects. Not only did he continue the experiments with Kulagina, but attracted a group of institute professors.

They had been working for several years, had obtained a lot of interesting data, which could be published only after many years. The main conclusion was that not heat energy, but some special impulse radiation came from the hands of Kulagina. This radiation caused telekinesis - movement of small objects - and in the most controlled conditions, changed physical properties of solid and liquid materials, and penetrated through electromagnetic screens. All these experiments were made at a highly professional level by specialists of the highest class 14. No reason to doubt the results. The death of Kulagina, suffering from serious contusions obtained during the Great Patriotic War, stopped the work.

But it was already clear for Prof. Dulnev that there was something outside the scope of the physical laws presently known to us. Some mysterious factor X. It manifested extremely seldom, only with particular gifted people, but it existed! Consequently, it would inevitably become a subject of research and, finally, would lead to the discovery of new laws of Nature. Therefore, when Prof. Dulnev had left the rector post, he started studying energy-information together with thermal physics. After a time he organized the Center of Energy-Informational Technologies and later on started investigating torsion fields.

We worked together with Prof. Dulnev for three years, and that was an interesting and productive time. Discussions with him always touched some new interesting subject. Particularly, Prof. Dulnev was the first in Russia who established a course of synergetics at the University and wrote a book on this topic. Students who were listening to this course took the progressive achievements of the Twentieth century science as a next subject, it didn't cause their surprise, and their main burning question was the exam. A young generation easily perceives new ideas and concepts; sorrows, victories and defeats of the minds which gave birth to these don't show up in the smooth lines of textbooks.

At the same time, the epoch of Prof. Dulnev affected his character; he is not always a plain man. "There is no room for two bears at the same time", as a Russian proverb says, - and later I passed to another department, keeping deep respect for Prof. Dulnev.

Thus Fortune made another twist, the spiral turned out to be above the previous point, and after many years I came back to the higher institution again, to the familiar environment of students, post graduates, academic councils, and slow scientific activity.

BREAKTHROUGH

"Two absolutely opposite sects attack me - scientists and ignoramuses.
Both laugh at me calling me "a dancing-master of frogs.
" But I know that I have discovered one of the greatest powers of nature."
Luigi Galvani, Italian physician (1737-1798)

I had a dream to create a computer version of Kirlian photography for many years. In order to do that, it was first of all necessary to input the glow images into a computer, and then process the images. I approached that aim several times: input the images by scanner, developed programs for calculation of parameter calculation, and tried to use huge television cameras. Something came out well (the analysis of after-death glow and EPC glow of liquids was based on computer processing); nothing was obtained with the TV camera - the resolution was too low and the image didn't resemble a photograph at all. At the same time, I was carefully following the development of the technology. One of the most significant criteria was the cost of components for the developed device - expensive technical toys were just beyond our means.

And so, in the middle of the 1990's the technology for home digital photography began to thrive. Matrices which previously had cost thousands of dollars, fell in price down to tens and hundreds. That made them accessible for implementation. We were actively discussing that issue with friends, checked possible variants, and in 1995 together with a group of colleagues we managed to create the first pre-production model of the instrument for the television registration of Kirlian images.

The instrument was yet quite incomplete, it consisted of a few blocks connected by wires, but most important was that it functioned. Blurred star-like images were obtained on a computer screen. That was victory! We managed to make the step for which I had been striving for many years!

The fact is that there are certain waves and flows in any scientific-technical line of investigation. If you get into the wave, you are carried along and new horizons are opened around. If you go against the stream, most wonderful achievements are lost, and unnoticed. How many interesting works published in good journals remained on the verge of science, attracting no attention and having no response. And that only because these works haven't got into the mainstream of scientific development. A lot of social and economic conditions are interlaced here, but this is the reality of our scientific world. Scientists who are capable of feeling these tendencies win the recognition; their works become popular and drive forward new scientific developments.

One of the main scientific and technical achievements in the world in the end of the Twentieth century was personal computerization. The new current in whose seriousness nobody had believed even in the early 1980's started thriving, created a powerful industry, and completely changed the face of the world. The Internet, which has connected these computers in a single world network, has become one of the greatest inventions in the history of Mankind: from the mass of separate states, divided by frontiers, distances and languages, the world has turned into one single space, where frontiers and barriers no longer exist.

One of the tendencies which appeared in the end of the Twentieth century, but was fully formed in the Twenty-first century was the transition from film to digital photography. Now we are already accustomed to digital TV and photo cameras, and in 10 years the film cameras will be the lot of professional athletes. Already now advantages of digital photography are indisputable, and only some technological and economical issues remain to be solved.

Kirlian photography was an interesting achievement, and in the hands of enthusiasts such as Dr. Mandel in Germany or Prof. Milhomens in Brazil it has given valuable information. But even after long years they haven't obtained real recognition. Spending time on solutions, development, and drying hundreds of photo prints was so inconvenient. And storing data was a huge problem! In Dr. Mandel's center, several rooms are filled with shelves with millions of photographs. Go and find something in such an archive! Let alone that it is very inconvenient to send such data to someone.

The modern stage is the world of computer files, electronic processing of information and digital images. The silver photograph becomes the lot of professional artists, and, having passed a twist of a spiral, returns to the bosom of art salons, as once it was at the dawn of its development. Gramophone records, video tape records and photographic films are inevitably replaced by electronic data carriers.

These tendencies, so clear to us today, were only outlined in the middle of the 1990's. But I felt them; therefore I was so glad at the creation of the first digital Kirlian camera. We started spending all available means for the improvement of that first model, and my wife Svetlana supported me in this work in every possible way, often denying herself the most necessary things. In half a year a model ready for practical work was created. Simultaneously we were developing the principles of image processing, together with the group of friends-programmers. The work was interesting, but it was totally uncertain if there was any practical meaning in all that.

Is there any diagnostic value in these computer blots? I started racking my brains over the puzzle of how to organize clinical tests, but here Providence intervened again.

"UPSIDE-DOWN" COUNTRY

One evening I heard a telephone call,

"Konstantin, hello, I'm a doctor from Australia. I came for a couple of weeks to my former native land and would like to meet with you. When could you find time? My name is Vagif."

Vagif turned out to be a dumpy, thick-set man about 45 years old, a native of one of the former republics of the Soviet Union. His life path reminded me of the first steps of Vadim Polyakov in many respects: being a technician, he took interest in massage, felt a gift of working with energies in himself, then he mastered acupuncture, treatment with herbs, and became a professional healer. In contrast to Vadim, at the first breaks of the Soviet stronghold, he slipped out in one of the chinks appearing in the former inviolable borders and put out to wander around Europe earning his living by healing. Running over the ocean of life, he was thrown to the Australian coast and got stuck there, clutching with claws at every ledge of the ground. In a few years of work he established a healing center "Aura & Body" in a good region of Melbourne.

"We got interested in your work on the Kirlian effect. We would like to invite you to work in Australia."

Those words were said when we first met were a total surprise to me. By then I had several times been invited to Germany, Italy, Finland, but those were short trips to Congresses. Moreover, Europe was not so far from us. Australia? This is the other end of the world! And how did they get to know my works?

"We will pay all the expenses and wish that you should bring your equipment. If you agree, as soon as I come home, I start getting the documents ready.

The reason for such straight persistency became clear only much later, already in Australia. However, I had agreed without hesitation: exotic countries, palms swaying above the sandy beaches had always evoked a spiritual response in me, and who would have refused going to Australia!

I had been in correspondence for half a year, and so I was standing before the window of the Australian embassy in Moscow. I was offered to walk for a couple of hours. I went out to a Moscow street and made my way in the direction of the "House of Books" on Arbat Street. And suddenly, having turned round the next corner I saw the building of the grand Temple in front of me. I saw it for the first time in my life and was astonished with the might of the strict proportional architecture. The rays of light were illuminating the gilded cupola, and the cross was shining on it, like a melted magma, radiating its own light. I became stockstill with admiration. This was the Temple of Christ the Savior, once blown up by Bolsheviks and renovated for public money as one of the first measures of the new Russian government. Much had been written in the press about that, but there is a big difference between reading about something and seeing it with your own eyes. I was looking at that sacred building, and a feeling was emerging in my soul that this was a special sign, not only the symbol of the rising Russia, but the symbol of the new stage in my own life.

Australia met me with the ocean wind, purple spectacles of sunset and the awaited for exotics: kangaroo, which one could touch, ostriches, and koalas. But there was no special time for exotics. Morning jogging, bathing, and then - to the Center to work. Vagif worked for 12-15 hours a day and expected the same from the others. He put the whole complex of the strongest effects on the patient: complete fasting during 2-3 weeks, energy massage, acupuncture, osteopathy, and individual courses of herbs. He was a perfect master of all these methods and spent many hours with every patient, developing strictly individual programs of treatment. It was a massive attack requiring serious effort from patients. But the results turned out to be amazing. Many diseases, which were considered to be hopeless in modern medicine, gave up before such an attack. Having entered an Altered State of Consciousness, Vagif could read information about the state of patient and thus direct the course of treatment.

I examined 10-15 patients a day, then, late at night we discussed the results with Vagif. After the very first week it became clear that my conclusions based on computer images of the glow from fingertips, with the help of a diagnostic table, correlated very well with the conclusions of Vagif and also with data of standard analysis. We spent long hours analyzing images and had no doubt in the value and convenience of this computer technique of analysis. Naturally, these were only the first steps, the most basic computer programs, but even those enabled us to make important conclusions on a patient's state and follow the dynamics of his treatment.

Holding our breath, we followed images of glow for a certain organ, influenced by Vagif by means of acupuncture or bioenergy, which were changing on the computer screen. The invisible world of virtual energies, controlling all the processes in the human body according to Oriental Medicine, appeared on the screen in the form of real images! That was exciting and thrilling! As if a whole new world opened up before us!

Sometimes at night we went out to the ocean shore and discussed the prospects for introducing our technology into the world of practice. Vagif offered to make that jointly, discussed that topic with rich patients, and got a promise of investments. Australia turned out to be a very fertile ground for the development of new ideas.

One day I finally got to know the real reason for his inviting me. There was a farmer in Australia, who claimed that he had been taken away by the extraterrestrials and had been shown around their planet. Who knows what they had needed it for, but the book written by him was quite popular in certain Australian circles. And so, when he had asked his greenish friends what new technologies would be on the Earth in the Twenty-first century, they had listed:

Development of gravitation,

Extraction of energy from vacuum, and

Diagnostics by the energy field.

Moreover, the latter, as stressed by the triangle alien manikins, had been based on the invention made in Russia long ago and had been developed by one odd Russian fellow. The farmer hadn't been an expert in technology and had never heard about Kirlian photography, at that. Those evidences had made an impression on Vagif, he had grasped the meaning right away, and found the one who had been working on the Kirlian effect through some Russian friends.

Three months in Melbourne contributed much to understand the analysis of images and, most important, instilled confidence in the value of the technique. There were very pleasant cases. Once a person visited my analysis at the Center. I found his field to be quite perfect, almost no defects, and congratulated him on that. After this conclusion he laid out a long wooden tube on the table - the national Australian instrument "didgeridoo" and presented it to me, explaining as follows. A month ago he had visited me and his field had been measured out of curiosity. I had told him that it had been possible to live in such a health state, but not really very good: his energy field had been unbalanced, the systems had been disharmonized, and all that had been manifested in a whole complex of symptoms: sleep disorder, anxiety, nervousness, tiredness, etc. This analysis had created a great impression upon him. Everything had represented the facts, but, as the majority of "practically healthy people" he had considered that it had been normal. But it had turned out that it hadn't been. And he had made a firm decision: had taken a vacation, had gone to the Australian desert, and had started to work on himself: exercises, diet, and philosophical novels. In a month he had felt himself another person, which was confirmed by our analysis. And it wasn't the only case.

Dark clouds slipped across the cloudless horizon only from time to time. Vagif was a real oriental despot by his nature, apparently some padishah in a previous incarnation. And those shah's habits were coming to light every now and then. When the employees of the Center, native Australians, showed disagreement with Vagif's decisions a couple of times, he dismissed them right away, having astonished the shy aborigines with his Caucasian temperament.

But "nobody's perfect!" When I was leaving the green continent on Christmas Eve, Vagif assured me of the soonest possible meeting and brought a pile of papers for the emigration of my family and myself to Australia. Everybody was excitedly discussing if there had been civil war in Russia and which mafia clan had taken power in the particular town, Chernigov, so the prospects of interesting work in the tranquil country were extremely alluring to me.

I was changing planes in the airport at Kuala-Lumpur on the way back. I had an hour between the flights and was slowly walking along the huge airport hall. A mixed multilingual crowd was all around me again; white, yellow, black, brown people. It seemed that all races and peoples were mixed in that Asian city. And suddenly a deep inner feeling, rather an inner vision, arose: I felt myself a part of this global international society, a part of this multinational mass of people whose name was Humanity. I had a feeling that my trip to Australia was only the first step in the long journey around countries and continents. And I would experience that coming to many more airports and cities, meeting different people and attending events in the future.

With that feeling I came back home, to my wife Sveta, children, friends, work, but was inwardly getting ready to continue moving. Winter passed and Spring came, Vagif made a contract with a fabrication plant for supplying a consignment of our instruments, but fewer and fewer words were about the emigration. By summer I understood that Vagif had given up that idea for some reason.

"Well, OK, we will survive here, but he could have at least told us about that", I shared my regrets with Sveta.

In summer we finally received an advance payment for 10 ordered instruments and started preparing production. Directors of the St. Petersburg company "ELSIS," with whom we were developing the optical block of the device, were discussing the prospective work with enthusiasm.

In July, together with Sveta and our small daughter we got ready for a trip to the Altai, and I announced to all that I would return in a month. The trip went according to schedule, with many fine memories and without any incidents. After three weeks we were again in St. Petersburg. And first of all I called "ELSIS"

"Hello, Victor, how are you doing?"

"Konstantin, is that you? Where from? "

"What do you mean "where from?" From home."

"But you should have returned only after another week!"

"I planned so. In reality I'm already here. How are we doing with the order? Let's meet in a day or two."

"Well, you know, in a day or two we can't. We are flying to Australia tomorrow."

"Where to?"

"To Australia. To Vagif. While you were away, he called and asked us to come as soon as possible."

I hung up. It was clear that it was impossible to prepare everything, to agree, and to make all the documents in three weeks. I realized that talks had been carried on behind my back for a long time. Everything was done with real oriental treachery. No way out. Vagif's reincarnation.

When the bosses came back in a week, we met at the plant.

"So, how was it?" I asked.

"Regards from Vagif. We signed an agreement for transferring to him the exclusive rights to world distribution of your, or rather our instrument. He found good investors, and we had a dinner with them in a real palace. What fine lobster was served there! "

"Congratulations, colleagues. And what is my role in your agreement? "

"You will be in charge of research and development. In future we will be able to pay you salary. And, if you collaborate with Vagif, he can appoint you the distributor in Russia. "

"Well, wonderful prospects. Signed is signed. You are better in that. We'll do everything to help you. Let us prepare the consignment for delivery." The plant bosses were obviously relieved. Apparently, they weren't expecting such malleability from my side. A week later 10 instruments flew to Australia. And a week later there was a telephone call.

"Dr. Korotkov, hello, we are from Australia. The instruments are showing wrong parameters. We can't work with them."

"Perhaps, the settings got maladjusted during transportation."

"But can you explain how to correct them? "

"I doubt whether I can. It's very complicated. It's art. Impossible to explain. Send them back. We'll adjust here."

"Vagif asks you to come to Australia for negotiations. He will pay everything. "

"Sorry, guys, but I can't. Too much work. "

The fact is that our team was making final tests, and we did our best to "specially" prepare the instruments. The software had been "mined" in case good reverse engineering programmers would have been found there. In a month after installation a big backdrop would have appeared in the screen: "EPC software demo version. Please, contact the developer." We didn't have an official contract; neither with Vagif, nor with the "ELSIS" plant.

And one month later we transferred all our production to another plant. All developments were in our hands. "ELSIS" was trying to outbid our engineers and programmers, but nobody entertained their suggestion.

This story continued in an interesting way. In the beginning of 2002 our guys found a web site, where "ELSIS" was advertising their device for taking digital Kirlian images. And a week later the directors of "ELSIS" called me and offered to meet.

I never recall offences from the past; therefore, I invited these colleagues to the institute with pleasure. They inspected everything, received our books as a present, and after a time we paid a visit to them. Their instrument turned out to be quite simple, at the level of our developments of 1996. After the visit we were exchanging the usual civilities when suddenly Victor said:

"Konstantin, you know, we have a joint patent with you, executed for our company."

"Yes, I'm aware of that." Indeed, while working jointly and having optimistic prospects we had started writing a patent application. Then our relationships had stopped, and only recently by chance I heard that there existed a patent and I was one of its authors. Rights for the patent belonged to the company, so officially I had no opportunity to claim anything.

"What do you think of paying us about 200 dollars for every instrument you produced?"

I was dumbfounded by such a question.

"Sorry, for what? Have you invested money in our development or we used some of your ideas? To my mind, you produce a device based on my ideas!" I replied.

"You are right to some extent, but we are the owners of the patent, and only this is important on the official level. So, think our proposal over. In a week we got an official letter with a threat of legal prosecution down to deprivation of property."

A patent agent easily settled this situation. The issue of that patent is described on page 88 in my book "The Kirlian effect," published one year before the patent application had been submitted, and, moreover, the application had been an exact copy of that. So, we can only wonder at the impudence of people, unable to give birth to new ideas, but trying to tear a small piece, if nothing else, from others.

This situation doesn't evoke anything except for a smile and compassion for people fussing over trifles in life. At the same time, I recall the days spent at the beach of the Pacific Ocean and the invaluable experience obtained while working at Vagif's "Aura & Body" Center with pleasure. And, in principle, it is good that I didn't stay there. I could have hardly been able to work under the despotic command of the reincarnated padishah. And Australia is a wonderful part of the world, but too far from everything else. It's not America.

LIFE REFLECTING THE EPOCH

Here I was going to tell some clever ideas about life in America, i.e. the USA, but before that I had decided to present the life story of my uncle, who has been living in America. Our life to a large extent is the life of those who surround us and whom we have been meeting in our life paths. The more so as fewer and fewer people who went through the Second World War, the witnesses of the past epoch, remain near us; therefore it is so important to keep their memories for the next generations. Here is one of the stories of that epoch, which I wrote down in the beginning of the 1990's from the words of my uncle.

We are sitting in his small apartment in Brooklyn, sipping beer; my uncle is leisurely telling his story.

"I finished school in June 1941, in Leningrad, Soviet Union. My last exam was on 19th of June, and the graduation party was on 21st of June. We gathered at school, joyful and nice: guys in shirts and ties, girls in festive attires. Congratulations, dances, and then all night long we were walking in noisy groups along the embankments of Leningrad. June is the high time of White Nights: the sun doesn't set at night, and it is possible to read a book easily in the street at midnight. And the beauty of the embankments of the Neva river during the White Nights is indescribable! There were unusually many people in the streets that night: graduation party in all the schools of the city is a big celebration, and well-dressed, cheerful people were dancing and singing to the guitar all over the city. Could have anyone of them supposed that for the majority of them that summer was the last one in their lives!

I came back home at 6 a.m. and went to bed right away. Suddenly I heard in sleep,

"Get up, Izya, get up! "

"What a bad luck", I thought, "I've just started to dream and somebody needs me!

I open my eyes - my mom is standing above me and shaking my shoulder,

"Get up, Izya, get up, the war has started!"

On June 22, 1941, Sunday, Hitler treacherously attacked the Soviet Union. All previous life with its problems, dreams, and plans turned into sweet memories in a blink.

I went to work as an apprentice of tailor and in a month I mastered the principles of this profession, which turned out to be quite useful for me in the future. Patriotic songs were broadcast, communist leaders made people sure that the war would be soon victoriously over.

However, the tone of their performances gradually changed: Hitler's troops were moving into the heart of Russia with confidence and in August approached Leningrad. After a brief discussion with the schoolmates, all our class filed applications to the army. As volunteers. Now I am the only one who survived.

It happened so that I started going to school one year later than everyone else, and in May 1941 I was 18. I was taller than average, the good Lord provided me with force, moreover, I was seriously doing heavy athletics during the last year at school, so I could bend an iron poker with my hand. Therefore, I was sent to the school of miners after being examined by the medical board. My mom, relatives, and my girlfriend were seeing me off. Everyone was crying. Nobody could imagine that I would see my mom again only in 30 years.

I got accustomed to the war conditions soon. Good luck. Our sergeant had participated both in the Russian civil war and in the Soviet - Finnish war. He had perfectly studied what should be done and how to survive. The first thing he started teaching us was how to survive in any conditions and come back to the base camp. Then the war separated us, and I didn't know what had happened to him, but I remembered his lessons very well. I will be thankful to him as long as I shall live.

I started as a sapper, and then I was transferred to a regimental scout. I was in the war till its very last day. Many a time I was in the enemy's rear; don't even remember how many prisoners for wounded interrogation I captured. Once we even took a fascist colonel when he was going from his field staff to the toilet in the yard. Thus we carried him away without trousers, through the back wall. I was decorated with the order of the 1st Degree Order of the Red Banner of Battle for that. There were also other awards: by the end of the war all my lieutenant tunic was covered with bars. Of course, I was several times but not seriously - just shot through.

When the war came to the end in 1945 our regiment was located in a small German town Koenigsberg. The first feeling was the sense of deep relief when on the 9th of May, 1945 complete victory over fascist Germany was declared. All's over, I've survived! Now - home, to Leningrad. During the assembly the commander congratulated us on the great victory and read the list of those who were put forward for a decoration. I was recommended for the rank of Hero of the Soviet Union. However, never received it.

Weeks were passing by: quiet duties at the regiment and joyful hopes of seeing relatives, mom, and girls soon. However, I could only guess who of them were alive then. In the evenings we gathered in the city bar and were drinking ourselves blind. At night batmen were driving officers home. But then something happened that totally changed all my life.

That night I was already heavily drunk. The bar was crowded with people - both officers and sergeants - noise, rumpus, hard to breathe in the tobacco smoke. Most of the people were standing - the room was obviously not intended for such a crowd of officers. I was sitting in an honorable seat - near the bar - I was much respected in the regiment. Then some lieutenant colonel, not one of our regimental guys, approached the bar. Immediately obvious - a staff drudge. White, well-groomed hands, glasses in fine frame. He took some vodka and looked around searching for a place. Not just that there was no place to sit, people had to stand back to back. So he turned to me and said:

"Hey, you, Jewish mug, give place to the superior, now!"

He said that, and the noise died away in the hall. Indeed, my appearance was typically Jewish: both the nose and curly hair. The Jews had always been oppressed in the Soviet Union, although officially friendship of nations had been declared. Anti-Semitism had been a hidden, but obligatory policy of all Soviet organizations. Being a Jew, it had been difficult to enter a university or get a good job. That was why I started working as a tailor in Leningrad. But the relationships in our regiment were perfect, and no one had ever mentioned my Jewish origin. That's why everyone got silent when this coxcomb said again,

"Hey, you, Jewish pig, give place to the superior, now. "

To say the truth, at that time I was hot-tempered. Moreover, drank a lot. And all the officers were looking at me. I got up slowly, without saying a word, scrutinized the man closely as if I wanted to remember him well, and gave him a belt on the face. But it was too hard. The lieutenant colonel gave a shot in the air, hit the bar and bumped into the shelf with bottles, so that they fell down like overripe apples. He fell on the floor and got silent.

One of our guys rushed to him, shook him, and then got up and said in a low voice,

"Izya, you've killed him! "

Clamor and rumpus set up. I was standing, perplexedly looking around. I didn't feel drunk anymore. One was to be shot, or at least sent to concentration camps in Siberia for that, according to the army laws. I went to the exit, and everyone stepped aside, as if I no longer belonged to that world.

I went into the street and stopped, as if in a fog. Our colonel - regiment commander running from the barracks approached me and said in a loud voice,

"Come back to your barrack, I will take duty officers and in 15 minutes we will come to arrest you. Remember, in 15 minutes."

Something clicked in my head. As if I came to my senses. I had perfectly developed the instinct to survive during the years of war. Thanks to the commander! He loved me very much!

I rushed home, pulled off the lieutenant uniform, put on an ordinary shirt, quilted jacket, put all my documents into a haversack, and 15 minutes later I was far away from the barracks, making my way through the dark night yards.

I was moving at night and catching up on sleep during the day. I had been a scout and could hide myself in any cellar, garret, or in the leaves of a big tree. Now there are many films with heroes like Stallone and Schwarzenegger, and I was like one of these.

After long consideration I decided to move to France. Somewhat farther away from the Soviet Union. Moreover, I was often told that with my curly hair I looked French. I got over to the South of France, where it was quiet, and found a synagogue in a small town. I was fluently speaking Yiddish, so that I could easily speak with the local rabbi. I showed my documents and told that I had to run from the communist regime, although I wasn't going into the details. I was even surprised with the rabbi's reaction. He burst into tears, embraced me, brought me to his house and offered to make myself comfortable. In some time, after a good bath, substantial dinner, and in clean clothes, I tumbled down and was fast asleep for 20 hours for the first time after many days. At the Saturday service the rabbi introduced me as a hero, having passed through all the war with fascists, and luckily the town mayor was an orthodox Jew. In a couple of months I obtained documents of a French resident and got fixed up in a job at a local tailor's. A couple of months later I was awarded with a French medal "For Courage" to my great surprise.

That strengthened my reputation once and for all and provided a stable attention from French women of different ages. It was clear that they were missing their men's caresses during the war. So, I seldom woke up in one and the same bed. And wine had always been in abundance in France.

Thus I had lived for several years, easily evading matrimonial traps. I could speak French fluently. But once I heard the news that the Jewish state was organized in Israel. "That is what I need," - I had no doubts in this thought for some reason. I told the rabbi about my decision. He promised to make inquiries and a couple of days later he informed that it was not advisable to go to Israel: European countries couldn't come to terms with the migration procedure of the Jews. England, always having strong connections with the Arabs, was mainly opposed to that.

"So, wait for some time, everything will settle down soon, and you will be able to go", said the rabbi.

But if some idea came to my mind, I needed to realize it immediately. Therefore, I declared my firm decision that I would be making my way to Israel myself. The Rabbi was trying to persuade me, but then he shed a few tears and said that he would prepare money for the trip in three days. Indeed, he visited some well-to-do citizens and gathered an impressive amount of money which proved very useful to me. A grandiose binge was organized, all my girls were crying, asking to go with me, and I promised each one that I would call her as soon as I got settled. Well, it turned out in a different way.

First I went to Marseilles, visited bars and spoke with seamen. After a lively discussion and drinking an immense amount of Port, everyone agreed that I needed to go to Sicily, where I would evidently find smugglers, who would bring me over the Adriatic. Thus it happened to be. Having spent two days in a bar in Palermo I found a captain for a reasonable price, who, as it came out, had already earned on the side by illegal transportation of Jewish migrants.

We were 30 on the ship. I had no idea what means the captain used to thread his way through the coast guard, but a couple of days later we saw the Land of Promise. We heaved to, and at night we made our way to the coast with no single light. No one of us was sleeping - all were standing by the board and peering into the darkness of the night. And then - the blinding light and a loud voice from a megaphone ordering us to stop. The captain ordered "Full speed ahead!" We started speeding up but then a cannon-shot rang out. That was an English destroyer, patrolling the frontiers of Israel. All of us were escorted to the board, interrogated and locked in the hold. The next day we were brought to the concentration camp in Cyprus.

I lived in that camp for 4 months. The coast of Cyprus was controlled by the English troops, so it was impossible to escape from the island, thus nobody was guarding us. The only duty was to visit morning roll-calls. It didn't turn out to be very easy - to come off from the soft breast of a warm Cypriote and walk to the roll-call. But I had to cope with that.

The island was the land of plenty - a real paradise on earth. The Englishmen could not compete with me neither in the quantity of the wine drunk, nor in the attention paid to me by pretty local girls. So I was even thinking about staying there a little longer. But then the camp commandant called us and announced that all who wished could be brought to Israel in English warships. As it turned out they signed some international agreement.

I was met as a hero in Israel. I was granted an apartment in the center of Tel Aviv and a good allowance for the beginning. And somehow right away I felt myself at home, among my people.

There were not many Russian Jews. Several times a week we gathered in a cozy restaurant and discussed scarce news, coming from the Soviet Union. We couldn't even dream about some postal mail or, even less, about telephone at that time. I found work, and life seemed to be getting into a routine.

But that didn't last for a long time. The Arab states attacked Israel. One month later after the beginning of the war two Israeli colonels came to me at my work and asked if I would join the army, as there were obviously not enough professional military men. I agreed without hesitation. I knew the art of war to a nicety.

I was given the rank of lieutenant colonel and sent to the front line, where I became a battalion commander. From the beginning I had to teach the guys everything, but they were learning quickly. Soon we already passed to the offensive and started actively moving deep into the Arabian territory. I was recommended for the title of "Hero of Israel" for a few successful operations.

But war is unpredictable. We were driving in a car, the day was calm, almost no shooting. And suddenly a mortar struck from the rocks close to the road. The first mine made havoc of the road on our way, I cried: "Jump out!" But we failed. The second mine hit the aim. Crash, flash, and I fell into darkness.

I came to consciousness 12 days later, as it turned out. At a hospital. I was the only one who had survived of the 9 people in the car. However, strongly shell-shocked. Perhaps, I wouldn't have survived if not for Ida, a young hospital nurse. She was holding me in her arms for many nights and, I suppose, instilled a piece of her life into me. As the doctor then said, I had no chances to survive after all my wounds. Nobody even hoped. You understand, no one has yet discovered medicines from death. Perhaps, only love can do something. So, she pulled me out with her love, didn't let me go, although I was already beyond the bounds. I even had visions, although I forgot everything then. A strange feeling when you are still here, but realize that you no longer belong to this world. So, if it had not been for Ida, we wouldn't be talking with you now.

After a time time I felt that I would live. Slowly I started regaining consciousness. But I was learning to walk for 3 months yet.

We married with Ida when I was discharged. I was near my thirties then, but it was time for me to steady down. And would you believe - I had been hunting for every skirt before, and then - as if cut off: we started living with Idochka, and I didn't even want to look at anybody else. Although the ladies were always hanging around.

I could hardly walk with a cane, so I couldn't serve in the army any longer. I got fixed up in a job again as a tailor in an atelier. We began to live well, we were not rich, but had a comfortable life. What was more, public honor and respect. Two daughters were born, began growing up to be pretty girls. Many years have passed this way.

Thus I would have lived in Israel, but once I received a big package by post. Letters from mom, sister, photographs, and an official paper with the Resolution of the Supreme Court of the Soviet Union on complete amnesty of civil crimes committed during the war. I looked at the photos, and my heart sank.

The Soviet Union had no official relations with Israel at that time and the situation was the same for a long period. The Soviet Jews were not released from the country. But we began to correspond and relatives were calling me home, describing their good life. Then I received an official letter in my name, with complete amnesty and reinstatement of all decorations. I was invited to come back to my homeland and was guaranteed good reception and attitude in a special letter.

We were discussing it with Ida for a long time, what and how. I wanted to see my relatives, to embrace my mom very much. Ida was born in Moscow, but as a matter of fact no one of her relatives was alive: they had got into fascist hands during evacuation and all had been killed in a concentration camp. After long consideration and doubts we decided to go.

It was not possible to visit the Soviet Union at that time, we had to go forever. We were traveling by ship, and there were tears in my eyes when we approached Leningrad. And I won't describe what happened when I saw my people.

We were received well. An apartment was granted right away, however, three times smaller than the one in Tel Aviv. We had lived and visited relatives and friends for two months; then I was offered a job in the central tailoring atelier. Good salary. But then everything started.

Each morning I came to work, and at 10 a.m. a black "Volga" car arrived and a mufti asked me to go for a talk. The director of the atelier saw us to the door, with a sweet smile and bowing. At that time one should not argue with the KGB. A few days later I told the director that I had no time to do anything at work, he replied that "the labor collective understood everything and would render all necessary assistance."

I was brought to Liteiniy prospect, into a huge grey building of the KGB administration, an office with Brezhnev's portrait on the wall. The owner of the office, a KGB colonel Alexey Petrovich, offered tea, cigarettes, and started a slow discussion. He warned me from the beginning that it wasn't an interrogation, they were just very interested in my impressions of life in Israel and Europe. I was free to tell anything I supposed to be necessary, from the very moment of escaping from the regiment. They were interested in the smallest details: who was wearing what, what they ate and drank, what they were talking about. I knew no special secrets; therefore, I was calmly recalling my life. It was even a bit of pleasure for me. When I passed on to the life in Israel, another colonel appeared in the room, the information had a more strategic character. Although, I could hardly tell anything new.

Thus it went on for almost two months. When my stories came to an end the visits stopped, but I still felt that I was under control. As a professional scout, I was used to constantly control the situation, and I noticed muftis watching me from a car or following me in the street several times. I had nothing to be afraid of: I had no contacts with imperialistic agents, sometimes I even played a trick on KGB men: I approached them and asked for a cigarette or asked to go for beer in the nearest store.

However, my Soviet relatives were quite nervous about my situation. A couple of times after feasts they sat with me in a corner one after another and started their stories in an undertone. Those were mainly stories about night arrests, awful weeks in KGB basements, violence and humiliation. One could get imprisoned because of coming late to work, reading English books, and making bold jokes. No causes were needed during Stalin's times, Brezhnev's times were a little easier, but the Siberian camps continued to be active. So they were afraid that my freedom was a temporary state. And, I guess, were also afraid for their safety. So, my arrival brought much hap piness to the relatives, but also much fear.

"Human-scoundrel gets used to everything", as it is said, and gradually the life got right. But Idochka, my wife, couldn't get used to that. She didn't like everything: long lines and empty shelves in the stores, when good meat could only be bought by knowing the right people, our tiny flat, anti-Semitism from most of the Russian population, suspicious glances of policemen in the streets. And after visits to the official institutions she came back in tears and was lying in bed sick all the day long. Officials knew how to professionally humiliate a person. That was part of the system.

Once newspapers and radio solemnly declared that Leonid Brezhnev had taken part in the international forum in Helsinki. No one of us paid attention to that, what does it matter what the Kremlin celestials were doing. But several months went by, and the Jews started telling one another in a whisper that the Soviet Union signed the agreement on free immigration of the Jews to Israel. In the beginning no one believed that, but the "Voice of America" radio started actively discussing this topic. Although more than a year had passed since these talks turned into a reality.

Once Ida comes from work (she worked as a female hairdresser), excited and pale, and says,

"Izya, today one client told me that a new committee on sending Jews to Israel has opened in Moscow. Izya, I want to go home, to the warm. I can't stay longer in this cold country." She said that and burst into tears. I tried this and that, and she repeated that she wanted home.

I took a compensatory holiday and went to Moscow. I came to the building of the Committee: there was a line of 1000 people and about the same amount of policemen. As if all the Jews of the Union decided to move to the West. Finally, it happened so.

The most part of the Jewish diaspora left the borders of the Soviet Union. And how many talented engineers, mathematicians, and musicians were there!

First time I came back from Moscow with no result. One should spend a week in that crowd on the street. I started asking relatives and friends, and after a time I learned that certain questionnaires could be passed to the committee. Three months later all my family was called to Moscow and then half a year later we were already stepping aboard the plane Moscow - Rome. Good-buy, Soviet Union, with all your friendship of nations!

There were no official relations between Moscow and Tel Aviv at that time, so all the emigrants to Israel were sent through Rome or Athens. We were met in Rome and brought by bus to the suburban camp for "Temporary settlers." It was spring time, everything was blossoming, birds singing, and freedom were in the air. Everybody was in good spirits; it seemed that we went out of a dismal basement to the bright sun. Life in Italy was wonderful: we were given pocket money and were traveling around Rome and outskirts with great pleasure. Ten days later we were invited to the committee and offered an alternative: to go either to Israel or to the USA. It was a great surprise to us. I was in the mood to go to Israel: friends and respect were there. But the girls unanimously cried to go to the USA. They were convincing me that that was the only chance in life, that the USA was the most developed and powerful state in the world, and that it was their old dream to go there. So, before a month had passed, we were coming down the ladder in the New York airport.

We were offered a choice of several places to live, and I chose the district of Brighton-Beach, where my friends had already been living. So, since then I have settled there. I haven't learned English well: Russian and Hebrew have been quite enough. But to tell the truth, every Thursday I'm brought to the school of English, and, moreover, I'm paid 50 bucks for that. In addition, it is a perfect place to meet with friends. My contused leg is getting worse and worse, so I do not walk farther than the bar around the corner. My girls got married, they are happy, have their own houses and children. They would have never got that in the Soviet Union. With Idochka we live well, although, to tell the truth, I felt perfect in any country. The most important is that I always had friends to talk with and to have a bottle. And vodka is the same all over the world.

With these words we took a quick one, and I made my way to the nearest store for the next bottle. I was walking along Brighton-Beach, thinking how life had changed over the last 20 years. In the 1970's America seemed to be a mysterious land, a country of Miracles to us, as to Columbus some 500 years ago, and now it is just another country of the European world, although very special, with its policy and mentality, but having no principal difference from the other world. A country, where it is a pleasure to travel, to do business and to discuss philosophical problems. However it would be better to do that in our home country, Russia.

AMERICA, AMERICA...

When we say "America" we usually mean the United States of America, although there are many other wonderful countries on the continent once discovered for Europe by Christopher Columbus. But the USA occupies a special place in the world community considering both the scope of its achievements and the level of its pretensions.

I am writing these lines in an airplane, on the way from New York to Amsterdam. The Stewardesses distribute wine and snacks, waters of the Atlantic Ocean being far below. End of October 2001, one and a half months after the attack on the World Trade Center.

This event will go down in history as the beginning of the new epoch of international relationships, new geopolitical redistribution. 10 years have passed after the communism-capitalism split was over, and a new world confrontation emerged. The world has been divided into two camps: the confederates of the USA and the enemies of the USA. Why USA, and not England or Germany?

I have happened to travel a lot around this country recently. By air and by car, it's a real pleasure to drive on their highways, until you reach a big city. There one can spend hours in traffic jams during rush-hours. Once I had to urgently get from Manhattan to JFK (John Fitzgerald Kennedy - international airport in New York), and the taxi-driver was watching a compass in his battered yellow cab making a big circuit, otherwise we could have spent several hours in a slowly moving flow of traffic on a central freeway in the middle of the day. The detour came out to be much quicker. The traffic problem has turned out to be one of the insoluble sticking points for modern civilization.

The American continent offers a variability of zones and landscapes and has probably no comparison with any other continent on the Earth. Unless Australia. Mountain-ridges with picturesque forests and lakes turn into rocky deserts, which gradually change to the blossoming valleys along ocean coasts. When you fly from the East to the West above the American deserts, you get astonished with the courage and will of the Americans setting off to develop new unknown lands in their wooden wagons. The lands where wild nature, hurricanes, fearless beasts, and ruthless Indians were waiting for them. The admixture of Indian blood is now considered to be a specialty, but how difficult it was for the first farmers to stand the raids of Indian warriors, fighting for the freedom of their prairies!

And these very conditions formed the national American character in many respects. Free, independent, relying only on themselves, generous, and pitiless. Until now this country doesn't forgive mistakes. You failed in business, you have chosen wrong partners, lost work - and everything is taken away from you, although a piece of bread for living is always left. America gives all the opportunities for development, but doesn't like unlucky fellows. This is a country where one needs to be an optimist and smile broadly. The Russian national hero is Ilya Murometz, who was idling away his time and "eating mother's potatoes" for 33 years, to become a warrior fighting all the enemies of Russia in later years. In the USA little uncles sams sell pastry on the streets from childhood and have their own business when they are 20. No one can afford to idle one's time there. The Americans are used to work, all of them are workaholics, and this attitude to life is being brought up since childhood. 8-10-hours working day, plus one hour for the way there and back - this is the order of life of a typical American. But at the same time, one can take a car on credit, buy a good house on credit, and doesn't have to wait for the old age to do that. If only they have work. And many families are wandering from one end of the country to the other searching for a good position and settling down at every new place, taking only a cat and family photographs with them. Well, if one has a good head on his shoulders and business intuition, he can be lucky to pass from the middle class to the class of middle millionaires. But it is not so easy to maintain oneself in that class. All my middle millionaire class friends work very hard, balancing in the conditions of fierce competition and dynamically changing markets. If once you fall behind, calm down, or miss new tendencies - then some other fast guy turns out to be ahead of you. And it is absolutely not easy to reach a stable multimillionaire level, when one can live in warm countries on a dividend income. And how many multimillionaires turned into common millionaires with the fall of stock prices after the end of Clinton's era! As in a well-known anecdote,

"Can a woman make a millionaire of a man? "

"Yes, if he was a billionaire before meeting her."

In Russia we live in freedom. If you have forgotten to pay for electricity - you'll have time to do that next month. Or in a month. Or in another month. Drive over the speed limit - deal with the inspector on the spot. He is happy, and you can drive further. Such things do not work in America. In America you must pay a fine to court.

But what is really impressive in the United States is the high level of professionalism in any question. Dilettanti do not survive there. Whether it concerns the development of new devices, Internet web pages, or arranging a lawn near one's house. If there is a task and a consumer, the solution will be given in the shortest time and at the highest level.

Another wonderful feature of Americans is their love of freedom and independence. The question of control of individuals by the state is constantly being discussed in the press. They are desperately afraid of being in the place of Soviet citizens, who had no rights, but only responsibilities. In Clinton's epoch the Congress passed the laws restricting FBI's rights of civil control.

For example, it was necessary to obtain a court's permission in order to tap more than one telephone of a certain citizen. Terrorists used that when preparing the September attack on the USA.

And, finally, there is no other country in the world, where emigrants would feel so free. A Russian in Germany will always feel himself an alien until the third generation. On the contrary, in the USA one needs to learn English a little bit, and becomes an American. Not loosing Russian background, at that. Indeed, this is a country where any newcomer feels at home. The country breeds a deep respect for individuals, regardless of nationality, habits, and color of skin.

Although, the last question - racial, still produces great friction in the USA. The difference between the white and the black population still shows up in many details. It doesn't exist at the legislative level though. Black citizens of the USA have open excess in all directions, until the level of the highest officials, and each president necessarily has African American representatives on his team, otherwise he looses the voice of a big electorate group. But in everyday life the distance between white and black Americans still remains even after the colossal progress achieved in the recent 30 years from the times of Martin Luther King.

Moreover, I have examples to compare with. The country where there are no racial limits at all is Brazil. White, black, yellow, brown - no one pays attention to that, all are mixed in a single nation, astonishing by happiness, optimism, and friendly attitude to one another. This is really a prototype of population of the Earth of the future, where all races and nationalities are mixed and live side by side freely and equally, maintaining peculiarities of their culture and traditions, at that.

And another important quality striking one's eyes in the USA is bringing up national patriotism and a sense of support. The September events caused shock among the population. But in some time the country came to its senses and united under the slogan "God save America!" The tragedy which occurred, a low attack on innocent people, then a large-scale operation of sending a bacterial weapon by post - that naturally changed the everyday business rhythm of the American society. Military operations on USA territory took place for the first time since the war between the North and the South. But after the first shock a back wave occurred - the wave of mutual support and patriotism.

The American flag was often seen before on public buildings and private houses, too. Now this tradition got a new powerful impulse. Not only buildings are adorned with American flags but many cars, as well. Badges with the American flag on suit lapels came to be usual and the President's call to unity came home to many people. Money from all over the country flew into the relief fund for families of the deceased. All well-known actors and public figures considered it necessary to take part in the acts of charity. The first shock and fright changed into the desire to act.

And this reaction manifests the main feature of the American national character. Deep individuality and independence, on the one hand, and capability to unite, to help a fellow when danger comes, on the other. The symbol of the American character today is a businessman-individualist, not letting anyone interfere into his business, feeling himself the Citizen of the Great Country and ready to fight for his own hand. Such a patriotic feeling was once also present in our Russia, and now all attempts are to be taken in order to bring it up again, especially for the young generation.

At the same time, the USA is far from being the land of promise. This country has many problems. And, first of all, problems at the level of individual life.

One of the advantages of the American way of life is a relative accessibility of good living conditions. Good cars, good houses. And that was achieved at the expense of introducing the most modern, cheap, and mass technologies. This approach fully proved itself and created a society of well-being in the sphere of material products. But when the same approach was used in the sphere of health, nutrition and culture, it led to losses.

Nowhere else in the world would you see so many deformed, shockingly fat people, as in the USA. According to the official statistical data, body mass of 30% of adults and 50% of the older population significantly exceeds the norm. And a whole class of diseases is connected with overweight. According to the data of medical statistics, high blood-pressure makes up 45% of all the diseases in the USA. Obese people are 85% of diabetics, and 35% of people with ischemic disease. The death-rate of the obese oncology patients is higher than the average of patients with normal body weight by 30-50%. For rectal cancer or prostate cancer in men and endometritis for women - the obese death rate is 5 times higher 15.

In the USA you get an impression that all the American people work for sugar industry. Coca-cola, pastry and sweet pop-corn are the national food of Americans. But the problem is not only sugar.

When the industrial production of chickens and bull calves started, these animals began to be necessarily fed with antibiotics and hormones, so that they did not get sick and grew quickly. And these drugs have been consumed by the American population for decades, together with hamburgers and chicken meat. As a result, many diseases characteristic of old age have suddenly "grown younger." Particularly, childhood diabetes, which was a very rare disease just some tens of years ago, is getting more and more widespread in the USA. It has recently been found that coronary arteries of 20% of American children and adolescents are affected by atherosclerotic plaques 38. The number of people suffering from various allergies is sharply increasing. More and more people are suffering from chronic skin diseases of unclear origin, starting from subacute inflammatory diseases of inner organs, joints, and musculoskeletal system to the so-called irritable bowel syndrome and other chronic diseases of the gastrointestinal tract. Such serious system disorders as the syndromes of chronic fatigue and seasonal depression, and the syndromes of childhood hyperactivity with attention deficit disorder have become a serious problem in the USA. Thus, the system of fast, cheap, conveyor food based on synthetic products created in the USA has caused great damage to the health of this population within a few decades. Of course, a patient is not left to the mercy of fate in such a developed country as the USA. One can expect that medicine will prolong his or her life for perhaps many years. But an increasing part of both private and national budgets is spent for supporting the vitality of the population. According to American medical statistics, public health costs were 989 billion dollars in 1995, and more than 1.3 trillion dollars in 2000. More than 14% of the gross national income of the USA is spent in medical industry 16. And most of these funds go to the multinational corporations, developing newer and newer (and more expensive) medicines. Doctors just can't keep up with the drug innovations, let alone the fact that it is impossible to predict the con sequences of taking several different medicines at the same time.

Recently many citizens of the USA have started have begun to realize the danger of "fast-food" products. Interest has increased in "organic" products, grown in natural conditions. At the same time, this problem has been almost ignored at the governmental level. Genetically modified products get more and more wide-spread, and, when consumed for a long period of time, the consequences are absolutely unpredictable. No one can foretell how that can affect the generation being born. In Europe clever biologists managed to achieve the prohibition of wide distribution of such products; but practically no one knows about that in the USA.

Quite recently subtle biochemical mechanisms have been discovered responsible for the transfer of heritable changes connected with the influence of the environment. This is the so-called modification variability. We are used to the fact that heritable information is transferred by genes, and many scientific works have been dedicated to the so-called mutational variability, determined by chemical changes in some or other structural genes. But this is an extremely rare process, taking place only under the influence of strong factors, such as radiation. It is impossible to transfer the acquired features or characteristics (for instance, a stable tan or ability to drive a car) from generation to generation in such a way. However, there are many examples of the so-called long-term modifications in the biological literature, when the characteristics caused by environmental influence can maintain for several generations. Long-term modifications are especially possible in cases when an environmental factor (temperature, low radiation, chemical elements of water and food) influence certain critical stages of embryonic and post-embryonic development.

Possible biochemical mechanisms for such types of transfer of hereditary information have been found in recent years. A special class of albumens has been discovered, the so-called "prions," which can change their structural-functional properties in response to the influence of environmental factors. Moreover, prions can induce similar changes in identical and related albumens, continuing even after the end of effects by the inducing factor. In principle, the discovery of prions demonstrates that heritable changes can be transferred not only by way of DNA, but also by albumen.

Thus, the reasons explaining why the number of children and teenagers suffering from "adult" diseases is increasing so drastically in the USA have been formed over decades. This is the back side of industrialization and the creation of the "society of prosperity." It is again worth emphasizing that the methods which have given brilliant results in technology have led to destructive after-effects in human health. Again and again we come to an understanding of the fact that the simpler life conditions are, the closer they are to the laws of Nature and the more useful for human health and longevity.

Fortunately, many people already realize the value of the system of individual healthy nutrition, individual water treatment, and an active life style and apply that to themselves and their children. Therefore, more and more people in Russia prefer to purchase natural apples and tomatoes, and not the picture-perfect synthetics.

An interest in questions of the scientific substantiation of matter, consciousness, and soul has appeared in the USA in recent decades. Formerly, a serious scientist could have lost their reputation, if they had started to discuss this topic. Now the topic has organically entered the scientific field of consciousness study 18. How do we perceive the surrounding world and ourselves in it? Is the brain the source of consciousness or just an actuating mechanism? From where does the religious perception originate? These and similar questions are becoming the topic of serious scientific research and discussions. And if the Americans take up some business, they do it thoroughly and develop deeply. I happened to participate in several scientific symposiums, dedicated to this topic, and every time there were a few serious, thoroughly done researches among the shared discussions. The newest series of experiments was connected with the application of computer brain tomography in order to reveal the neurophysiological processes concerned with the religious perception. Having investigated the brain functioning of monks during prayer and yogis during meditation, the authors came to the conclusion that certain zones in the cortex of the cerebral hemispheres, connected with one or another feeling (feeling of flight, ecstasy, inner voices), can be identified. But this observation would be too simple to explain the whole range of spiritual feelings and emotions, the feeling of divine grace and happiness. Other methods and many experiments are required. Our EPC bioelectrography technique turned out to be quite useful. But first of all, let us dwell upon this very technique.

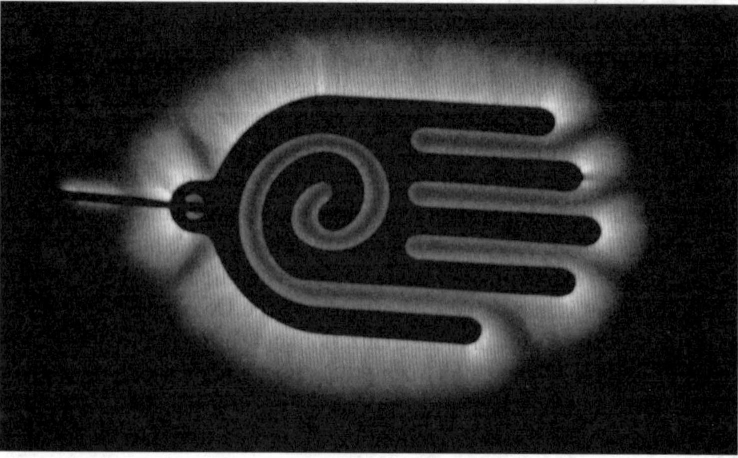

BASICS OF EPC (GDV) BIOELECTROGRAPHY

Spring 2003, EPC workshop in San Francisco, about 50 people gathered: physicians, scientists, and those who are interested in new scientific trends. This is quite a big number by American standards. After the first day of theory and lectures, we passed to practical demonstrations on Sunday. First, as usual, registration of the field picture of one of the volunteers, doing exercises and monitoring the effect on psycho-physiological state. A nervous lady, around 55, volunteers. We measured her field from fingers. The picture was awful, the stress level was 6.8 with the norm from 2 to 4, although the physical state is quite good. She performed respiratory exercises for 5 minutes. We measured the state again - it was even worse. I wondered, how she could teach courses on self-regulation having such a state?

A second volunteer - a software engineer Bob, a friendly smiling gentlemen of stout constitution. We measured the initial state, after which he concentrated and began meditating. The curve of state moved confidently up. We repeated, and the result is the same, the change of state is statistically reliable and reproducible, although, judging by the character of the curve, it demonstrated concentration rather than meditation. Well done, Bob!

After lunch I suggested we made a small experiment. I used the instrument to measure the dynamics of gas discharge glow of a sample of drinking water. The picture looked like a small circle with a "tail." The time dynamics represented a curve with small oscillations. It was typical of drinking water.

The instrument left untouched, I suggested all the participants to meditate based on the idea of the movement of energy flows. One needed to relax, close his eyes, imagine how the energy fluxes move in the body, coming up from the feet, through the body, and dissolving in the head. After some 10 minutes of meditation I asked everyone to concentrate on the water sample and, sending positive emotions, turn water into wine. Another 5 minutes of concentration.

(I organize similar experiments at practically every workshop in different countries. The main condition of success - the confidence of people in a positive result and a deep inner concentration. One needs to believe in success, as in any business one undertakes. Moreover, it is very important that the whole group works in unison. One Skeptic, if present, can spoil the whole result.)

I measured the water again. The difference was seen without any devices. The glow picture turned into a pentagonal star, variations of the dynamic curve sharply decreased. This is quite a rare result. Usually quantitative parameters changed, but the character of the image remained the same; here the change was very significant.

We repeated such experiment many times; the probability of success depended on the group in many respects, on people's ability to create a special state and concentrate the attention on the instrument with the water sample. Naturally, we always perform control experiments, when the water stays without any influence. No change takes place at that. These experiments are coming along with the work performed in the laboratories of different countries and show that human Consciousness is the moving force in the world surrounding us. Based on the results obtained, we decided to start a large international program, performing the experiments with the registration of distant influence of humans on water. From Germany to Russia, from Russia to the USA, from Japan to England. Registration shall be carried out both by our instruments and standard methods. After 1-2 years of work it will be possible to generalize all data and perform a so-called meta-analysis, i.e. compare the results obtained by different researchers in different countries of the world by various devices. The main emphasis will of course be laid on the EPC camera - our most well known development.

We have managed to progress a long way since the creation of the first model of a computerized Kirlian camera. A principally new class of equipment has been created, based on software processing of gas discharge images and extracting information on the object's state, whether water, homeopathic solution, grain, or human being. We have managed to demonstrate that the glow around the object arises as a result of emission of photons and electrons from the object's surface under the influence of the electrical field impulse. It is as if the field draws particles out of the object, stored at the expense of free energy, and amplifies them in gaseous discharge, like the process in photoelectron multipliers and counters of radioactive particles. Therefore, we began to call the technique "the Gas Discharge Visualization technique (EPC)." Graphically speaking, this is the digital computer stage of the Kirlian effect. They are correlated as a bicycle with a "Mercedes." Both are vehicles, but the result is a little bit different.

Our research demonstrated that, first of all, one should forget "kitchen" instruments and pass to the professional level. A specific form of impulses and stability of all the parameters, not less than 1%, are required. And, naturally, techniques for image processing are needed. Today's computers enable to perform the processing which was only dreamt about in the past. But still the same Kirlian effect lies at the basis discovered more than 200 years ago by George Lichtenberg. One of the main conclusions recently achieved is that the EPC information is obtained at the quantum level. Applying an electrical field, we stimulate quantum processes: photon and electron emission from the object's surface, for example from skin. And the entire organism as a whole turns out to be involved in these processes. Therefore, the approach pertains to the new directions of quantum informational biophysics. Why informational? Because the analysis of images is performed on the basis of modern methods of information theory, particularly, non-linear mathematics 19. Within a couple of years a whole complex of software was created, enabling us to represent, in various graphic forms, the parameters calculated on the basis of EPC images. Appearance of different program screens are shown in the pictures. When the first

EPC cameras were created, the main attention was paid to the scientific examination of the received data. How can it be used for an evaluation of the patient's state? What comes out of the analysis of psycho-emotional state? Are there some peculiarities of this glow of water and blood? A whole number of Universities and institutes took part in the research. Scientists of the highest class participated in the work, international scientific projects in Sweden, Finland, Germany, France, and USA were started. The results came out to be so interesting that all who started studying the EPC technique became our active followers in some time. The method turned out to be sensitive to the very subtle changes of state of the studied object, be it a patient after an acupuncture session, a sportsman during training for competition, or homeopathic solutions in different concentration. The four main spheres for the application of EPC technique were found: medicine; consciousness study; sport; liquids, materials, medicines.

Particular fields of application can be distinguished in all these spheres, giving the most interesting results.

The EPC Camera becomes more and more wide spread in medicine. After long-term clinical tests the Ministry of Health of Russia approved it as a medical instrument and included it in the catalogue of medical devices. Three dissertations for the Ph.D. in medicine were defended on the basis of the obtained data. We hope that in the future the EPC Camera will be used in every clinic along with X-ray and ultrasound devices.

Imagine: you come to a clinic, and the first thing you do is have your fingertips measured. Why the very fingertips, and not some other part of the body? Information on the organism state can be obtained from various parts of the body, and we have a special scanner designed for observing the glow of acupuncture points and meridians. (It is worth mentioning that such instrument was first developed by the Kirlians, and then reproduced by Romen Abakyan - the Armenian physicist, working in Moscow). Fingers are very convenient objects for investigation. Most people have them, and the absence of one or two fingers does not prevent making an analysis. A patient need not be undressed, and the research can be performed in any available room.

In some minutes print-outs are passed to the doctor, with an analysis of the functional state of the main systems of the organism, their interrelation and potentially dangerous zones. What is more, a quantitative parameter characterizing the level of stress is determined. Now the doctor can decide what other more detailed tests should be done with the patient and what specialists shall be contacted, i.e. that is the first stage of express-analysis, objectively characterizing a patient's psychological and physical state.

Naturally, this is not a magic mirror. As any analytical method, EPC bioelectrography has its limits. But if used in skillful hands, it provides most valuable information.

One of the first researches of gas discharge parameters on biological samples was initiated by Professor E.I. Slepyan. Once he called me and said,

"Dr. Korotkov, can your technique be used for the investigation of quality of meat products, for example, the degree of meat freshness?"

"Why not, Dr. Slepyan, we can try."

We discussed the technique for setting the experiment, Prof. Slepyan gave samples, and in several days we received graphs, showing reliable difference of various samples of beef. Prof. Slepyan was very glad, he took the materials and called me back almost the next day,

"Dr. Korotkov, I got in touch with the director of our meat packing plant, he wants us to discuss your data with our meat quality laboratory employees. Could you come together with me Wednesday morning?"

I agreed, and in due time we were received by nice ladies in white smocks who first of all gave us tea. A wide variety of the most delicate samples of the plant's production was offered with the tea. I was glad again that I am not a vegetarian. When we finished at the table and tried to pass to the topic of our visit, the nice ladies explained to us that they liked us as scientists and gentlemen, but our methods were not at all interesting for them. The meat could be smelled and touched, and if there were some suspicions, it could be processed and used for sausage. And no problem, at that. Naturally, that was far before the epidemic of the cow rabies disease. Later together with Prof. Slepyan we were carrying out a whole series of projects on water, plants, materials, somehow connected with ecology. And now the next co-authorship article is being prepared. Water has been our favorite object for a long time. This amazing liquid, the basis of life on Earth, offers new developments to the scientific world. It was found that apart from the three well known states: liquid, solid, and gaseous, water can form groups of several tens of water molecules characterized by different orders of degree. The so-called clusters are formed - groups of water molecules making up a stable geometrical configuration. The formation of these clusters depends on the external conditions of water. Say, on the influence of a field, for instance a magnetic field, or the geometry of surrounding space, or on the presence of extra-low doses of some substance. When clusters are present, we obtain structured water, and this structure can be maintained over a long period of time. Thus, we can discuss the memory of water. All these notions, previously waved aside with depreciation by the academic scientists, have lately become a subject of serious scientific discussions. First of all, because the experimental techniques appeared enabling us to investigate such extra-weak effects. Moreover, there are no doubts that many of the extra-weak therapy techniques exert a significant influence upon human health. One of the methods of investigating "informational" effects is the EPC bioelectrography. It has been more and more widely used for the study of homeopathic solutions, 20 natural and synthetic substances, 20 and drinking waters of different origin. A difference between blood glow of healthy and oncological patients has been found. What is more, comparative simplicity and availability of this method has opened broad prospects for its application in practice. Exotic variants of applications appear. It was shown that the glow of water changes after a night spent in a model of Cheops pyramid. Such experiments have several times been reproduced both in Russia and in the USA. Interesting, that the effectiveness of this influence varied greatly, and that variation was connected with uncertain factors. Being carried away by this topic, we went to the country of pyramids - Mexico - and registered the influence of pyramids Teotihuacan and Chichen-Itza. Then we moved to Peru, Brazil, Korea, and, everywhere, we investigated the influence of energy active zones on the water glow. This topic is very interesting, and now a big international project on the research of EPC parameters of liquids is being prepared. The glow of a water sample placed on the electrode of the instrument changes under the influence of the consciousness of a human, although not every human. The collective influence of a group of people works very effectively. Say, when all the group participants try to influence the water sample with their positive feelings and emotions. Naturally, at a distance. We have been regularly carrying out such experiments at our workshops. We measure the glow of water in the initial state several times using our EPC instrument. We demonstrate that the parameters of this glow are reproducible and stable. Then we ask the workshop participants to start the meditation and direct their positive emotions and, first of all, their love, to the water sample. When we repeat the measurement in 10 minutes, the glow parameters change. Isn't that the best proof of the fact that our Consciousness can directly influence objects of the material world! Such experiments enable us to understand why a home dinner cooked with love is much tastier than a dinner at a restaurant. The influence of biologically active zones can be studied in the same way. Although, special sensors can be used for that purpose. Thus, the sphere of application of the EPC technique is being more and more expanded. All the directions developed in recent years can be united by the notion "study of energy fields." Certainly, this approach is just beginning to be actively developed, and a lot of new methods of experimental research will appear in the course of time. Convincing data can be provided only when several different techniques are combined. Our Russian science appeared to be the most advanced in this sphere. This is first of all provided by the historical tradition: Prof. A.G. Gurvich was the first one in the world to create the theory of a biological field and experimentally prove it. This direction was further developed by P.I. Gulyaev, A.P. Dubrov, V.P. Kaznacheev, A.S. Presman, U.A. Holodov, and many other researchers. It has never been a priority in European science and has mainly remained to be the lot of interested enthusiasts. And in Russia we can afford to do something of interest. To develop one's ideas, even the stupid ones. As compared to Western scientists, we are freer. We are not restricted by a high salary and the necessity to get it. Many people think that in the West the researchers work in perfect conditions and get a good salary for that.

It is so, if these researchers develop in the acknowledged directions, especially in applied spheres. Salaries of scientists working for companies are several times higher than of those at the universities. But in all cases the scientific topics must agree with the general directions. Minor lines of research are not welcomed and not funded, and often are even persecuted. There are quite a few of such cases in the history of American science, for example, complementary medicine. So, it happens that the main scientific lines of research in the West receive investments running into millions, and the non-orthodox ideas must force their way through.

One of these fields turned out to be complementary medicine, making use of the non-pharmaceutical, extra-weak effects on a patient's organism: homeopathy, herbs, massage, acupuncture, and physical fields. This approach was neglected and suppressed, mainly by pharmaceutical companies and the American Medical Association. It has recently been objectively shown that allopathic medicine turns out to be helpless in the face of many chronic diseases, while weak task-oriented interventions can relieve some chronic conditions. The mechanism of effect from such influences is not yet clear in many respects. A wave of public and professional interest in unconventional medicine has risen in the world. And here it has been found that the EPC bioelectrography technique is one of a few methods enabling us to objectively observe the effects of informational influences.

In recent years I have been able to give talks at Congresses and Universities in dozens of countries around the world; in practically all European capitals, in many cities of America, both North and South, in the countries of Asia. And very often after the presentations, people come and express their admiration of the level of development of Russian science. And at such moments I feel myself a plenipotentiary of Russia, representing not only new scientific and technical achievements, but creating the notion that Russia today is not only a chaos and economical problems, but a colossal intellectual potential, playing by no means the least part in the modern world.

It is also important that Russia's one foot is in Europe, and the other - in Asia. Therefore, the Russian intellectuals have always been open for the knowledge of Eastern ideas and their transformation into the European reality. And in our work we attach significance to that and broadly use the ideas of both Chinese and Indian traditional medicine. And I consider that the idea once given by the outstanding physicist Erwin Schrodinger in his time is very important, "Still, it must be said that to Western thought this doctrine has little appeal, it is unpalatable, it is dubbed fantastic, unscientific. Well, so it is, because our science - Greek science - is based on objectivation, whereby it has cut itself off from an adequate understanding of the Subjective Cognizance, of the mind. But I do believe that this is precisely the point where our present way of thinking does need to be amended, perhaps by a bit of blood-infusion from Eastern thought. That will not be easy, we must beware of blunders - blood-transfusion always needs great precaution to prevent clotting. We do not wish to lose the logical precision that our scientific thought has reached, and that is unparalleled anywhere at any epoch."

Such attitude can help to save the lucidity of thought, a soberness of mind, and avoid falling into cheap mysticism, so popular in certain circles nowadays. One shall strive for balance in everything - both in relations, and in practical business. Although, it is usually not easy. And here I would like to give an example of such a balance - short essays on some of our latest expeditions.

WORLD WITHOUT BLINDNESS

For many decades in Russia there were people known to claim ability to distinguish colors and even read texts without using their eyes - by scanning the material with hand or applying it to the temple area. In the Soviet Union, at the beginning, it was considered as fake and a lot of efforts were taken by the authorities to disprove these abilities. Several people were even put to jail under forged accusations - the Soviet materialistic world-view was unable to accept unusual psychic abilities. With time there appeared more and more people demonstrating "skin vision", and the phenomenon was officially accepted, but research was not allowed, results were never widely proclaimed and only several articles were published in special journals.

After the collapse of Soviet Empire people obtained a freedom to follow their interests and present their results to the public. Several groups were involved by that time in the "skin vision" training, but most of them did not survive difficult post-communist times. The most effective was Dr. Vyacheslav Bronnikov, who has developed a system of training, which he coined as "direct vision".

Dr. Bronnikov approached me in 2003 and offered to conduct experiments with children trained with his technique. One day I came to his training center in Saint Petersburg and was really impressed by the presentation. Just imagine: boys and girls of different age - from 8 to 20 years old, tied tight black bandages around their heads to cover the eyes and without any visible difficulties read books, played games and operated in the environment. Of course, I tried their bandages - and was unable to see anything - they were specially designed to prevent any external light.

Dr. Bronnikov 's children passed a lot of tests - the first accusation was that they have special holes in their bandages. But all the tests conducted in most prestigious academic institutions proved that children could really see without their eyes.

At the same time it became clear that they comprehend visual information - when we placed non-transparent screen before the text, or conducted experiments in darkness, they were unable to see.

The main hypothesis agreed by different researches was that children comprehend visual information by means of their skin, and brain may process this information converting it to images. Children were trained to create a "mental screen" and project information as if an "inner movie". This approach was based on ancient trainings of some Tibetan monks.

Together with Dr. Pavel Bundzen for several years we conducted experiments with children of Dr. Bronnikov. They were seated before the computer, blind-folded; we had all our equipment around, taking measurements from their fingers, measuring the time of their reactions, while projecting to the computer screen different pictures: a man, a horse, a tree, a building, and so on. Children had to tell, what they saw on the screen, and we measured the time and dynamics of their reactions. The aim of the study was to see the peculiarities of their psycho-physiological reactions.

All children were supervised by the assistant of Dr. Bronnikov - Lubov Lognikova. She was a nice soft-mannered lady in her forties totally devoted to her boss and to her children. She was very attentive to every child and always tried to make all of them feel most comfortable.

Our experiments were very interesting and generated a lot of experimental data. After a time it became evident that in the process of "direct vision" children transformed to the Altered State of Consciousness (ASC), the condition, where brain operates in the most effective way producing sometimes amazing effects. This condition is typical for experienced meditators, creative people at their best moments, chi-gong and yoga masters. After many years of study we were able to find specific correlates for the Altered State of Consciousness, and we presented our findings both in Russian and international scientific journals[32,33].

But, of course, our research laid only the first stones in understanding the enigmatic process of direct vision. We had much more questions than we had answered. So we discussed the next experimental sessions and decided to have them done in the nearest future. I visited the training center of Dr. Bronnikov in a little Ukrainian city Feodosia, originated from Greek times, and located on the coast of the Black Sea. Dr. Bronnikov was able to build impressive facilities, where every summer he and his team hosted groups of people interested to get training. He offered training to people of any age, but only children were able to get direct vision abilities. Actually, not all of them. Dr. Bronnikov's training took a lot of time; children had to have everyday exercises, so it took a lot of concentration both from children and from their parents. And sometimes it took years before children were able to see. Many times I asked Bronnikov about statistical effectiveness of his training, but never had I got an answer.

Most of his pupils were children with bad vision. To them training was the most effective: they had a great desire to see the world in full. Blind children, who had lost their vision at some age, were quite effective as well. For children having no vision from birth it was the most difficult - they had no understanding what it means - to see. At the same time Bronnikov had several successful cases of blind children who were able to get direct vision ability, to begin normal life and after finishing school to enter a University. But even for children who were unable to see without their eyes after the course of training it was very useful, as it allowed them to be more effective in school and life.

It had a similar effect on adults. Bronnikov claimed some successful cases, but never delivered information about the people. I came to the conclusion that this is pretence but not the reality. He himself was unable to reproduce this effect, as well as his close followers. They were able to train, but were unable to see. Later this effect was confirmed and it became clear that we deal with a specific phenomenon.

Unfortunately, our research with Bronnikov was sadly interrupted. Unexpectedly, professor Bundzen died from stroke, being only 68 years old. It was a big loss for us all and we miss him tremendously - he was not only a brilliant scientist, but an understanding and kindhearted person. Some time later Lubov Lognikova developed cancer and died within two months. Without her careful attention the research process has become practically impossible - Dr. Bronnikov was too busy with his everyday responsibilities to find time for the research. So I decided that we should close this project and put it on a shelf. But life is preparing its own surprises.

By mid 2005 being in the USA I received a phone call.

'Hi, this is Mark from New York. We have a holistic center in Brooklyn. Sam was working with us and strongly recommended us to meet. How can we do this while you are in the US?'

Sam is a good friend of mine, doctor of medicine, psychiatrist and a talented healer. He can see energy fields, work with different healing modalities and clinically evaluate the treatment process. Sam's recommendation was significant to me, so after some consideration I answered:

'OK, it happened so that I have my ticket to Russia with a stopover in New York for 7 hours. We can meet at the airport'.

'We will meet you at the gate, take to our center and then bring back for the departure. Thank you, professor'.

It is interesting, that I had no special intensions to spend time in NY, but it was the most convenient connection that provided my flying from coast to coast of the United States. So one morning I was met at the gate in JFK airport by three Russian people. While we drove to Brooklyn they introduced me to their center and their activity. After describing treatment modalities typical for the holistic center one of them said,

'My specific interest is teaching people to see without eyes. I have a course in Brooklyn, but only for Russian people - we still have language problems.'

'Very interesting, are you a pupil of Dr. Bronnikov?'

'No, I've heard of him, but never met. I came to my approach myself, having an inspiration one day. My background is engineering, I was raised up and educated in Moscow; I was a very materialistic-minded person, far away from different metaphysical topics. Then one day, after having strong stress, I found out that I can read information on the past events. I can not see situations in the past, but I KNOW how it was. I began reading books, just by that time we decided to move from Russia to USA with my family, and after a lot of different situations and life experiences I established training course to tell people about subtle energies, about their hidden abilities to accept information. Eventually it became Mark Komissarov's training center. Children are trained to see without using their eyes; adults are trained to develop their intuition, to accept information from the Universal Informational Database.'

At that moment we arrived at Brooklyn and our car stopped at the corner of Kings Highway and 18th Street. The Center occupied several rooms of a two-storied building sharing it with a dentist's office. After a traditional cup of tea Mark said,

'Let me show you how it works. For your visit we invited several children for a short demonstration. They are waiting for us upstairs'.

We entered a room, where a boy of 10 and three 8-9 year old girls were waiting for us.

'Hello, my dear', greeted them Mark in Russian, 'Glad to see you. Let me introduce you to a professor from Saint Petersburg. He is a scientist and we'll show him our exercises. By the way', Mark addressed to me, 'Three children attended my training before, and this pretty girl - what is your name - Gita, thank you, - is for the first time with us. We are glad to have you with us, Gita. Let us show what we can do.'

Three children produced black masks made of thick fabric, and covered their eyes. (Later I checked these masks - they were absolutely non-transparent, covering part of the face). Mark presented them a thick book in English, and they one by one were sight-reading lines from the book, on the pages chosen by me. It was clear that they could read the text from the book. I turned the book upside-down and a girl read the sentence. After this Mark placed on the table plastic cups of different colors, and children collected on demand red cups, blue cups and then the yellow ones. A similar exercise was repeated with color sheets of paper. After this they played with a ball throwing it to each other (still in masks). All this was very impressive.

'But how long does it take to teach a child to see blind-folded?' I asked, 'And what is the effectiveness of your training?'

'Effectiveness depends on age', responded Mark, 'For 4-12 year old children the effectiveness is about 99%. In our practice one boy out of 100 being trained refused to follow the instructions. For 12-18 -year old the effectiveness is about 20%, while after 18 it is practically impossible to see without eyes, but people may be trained to access information directly. They do not see but they know. For example, one lady had to pass driving license exams. She did not try to recall the right answers - she just marked them one-by-one from the offered answers - the scores were about 70% and she passed the exams.'

'But this way it is possible to win the casino', I exclaimed.

'It is possible. I had one pupil who did it this way. After every night at casino he had a positive balance. But after some time of winning he was advised by security to leave the casino and to never attempt coming back again. He tried another casino in another city, but half-an-hour after he'd started gambling the security approached him and asked to leave. It became clear that his picture was added to the global "Casino Black List"'.

'But this is against the law! How may they forbid people to visit any place! The USA is a democratic free country', I protested.

'All casinos and clubs are private companies. In their regulations it is written that they may decline any customer without explanations. Seems to me, they may have bad cases of "lucky hand" before. This is allowed only randomly - to support the legend. But let us come back to our training. This girl - Gita - is here for the first time. Let me take her to another room and I will give her the first training'.

They left the room and I asked the children questions. They all had normal vision so they didn't need "direct vision" in their everyday life. But this ability helped them a lot: for them it became much easier to concentrate attention and their school marks improved significantly. Demonstrating their abilities to schoolmates they increased their social rank that allowed them to get higher self-esteem. So both children and their parents were highly satisfied with Mark's school. They conveyed this message to their friends. So Mark always had pupils even in a not-so-big NY Russian community.

While we were discussing, Mark and Gita came back. About 15-20 minutes passed. Mark put a black mask on the girl's face, opened a big book and pointed to a picture, 'What is it?'

After several seconds girl answered, 'This is a cow, black cow'.

I grabbed a book from the shelf and opened a page with a colorful picture, 'What is it here?'

The girl slowly answered, 'A house, green; a tree nearby, and two people'.

She was facing the picture with her left temple and answered slowly, as if revealing subjects in the picture one-by-one.

I was shocked. 15 minutes of training and - "direct vision!" How is it possible? Bronnikov takes months to get it, with a lot of exercises and much less effectiveness.

'Mark, how do you do this?' I exclaimed.

'This is my own technology. Next time when you are in NY we may discuss it in details. But direct vision is not their only abilities. They may read information from the past. I am not interested in "reincarnation memories" - our children watch cartoons and movies from birthday - I check it in simple experiments. Let us try it'.

Mark asked children to leave the room. Then he placed several objects on the table: a comb, a pen, a box of matches. I added a little Russian icon and a little doll from my pockets. The objects were on the table for several minutes, then he removed them and the table surface was empty. Mark invited four children to the room and they approached the table.

'OK, my dear', Mark announced; 'Now you need to guess what was on the table and write it on a peace of paper. Please, do not help each other. Just imagine how it was several minutes before'.

In a couple of minutes he collected the papers. Boy hit all 6 subjects, girls got 4 or 5. Gita did not participate - too early - explained Mark. All spotted "a bright gold picture" and a doll. It was clear they did not know beforehand which subjects I would produce and the only possible accusation in deception from Mark's side would be using telepathy. It would not be that bad either! But Mark did not pretend it.

'In our system they are trained to access information, be it in the present or in the past. This is more difficult than direct vision (sic!) and depends on individual abilities. Most may guess 60 -70%, some do remarkable things. I had two pupils who were able to read information from computer discs keeping them in hands. It was really amazing. That means, they had access to coded information and the disc was just the connecting link. But this is quite rare and needs a special deep concentration. So after some time one girl got tired and lost this ability. She did not need it in her everyday life'.

By the end of the session I offered to take reading of Energy Fields from children. Their EPC images with filter were very good, which indicated good health state. Images without filter were in the deficiency area with a strong left-right misbalance for all of them. This was not surprising - it often happens for children of 8-11 years old characterizing their highly sensitive and vulnerable energy system. Then I asked them to put on masks and watch pictures from my computer screen while I was taking readings from their ring finger. They were able to correctly distinguish pictures of people, cars and animals but when I produced landscapes, where there were no clear details / (with no clear details) they were able to tell only the color. The registered patterns of EPC time dynamics were very similar to those in Bronnikov's data.

My time was over - I had to depart for the plain. Soon our Boing-757 was high above the Atlantics and I was still under the impression of my visit. 1000 questions circulated in my mind.

The first hypothesis was that children comprehended visual information with their skin, and during this process the brain was able to create new connections between the skin sensory area and the visual cortex area. These neuron connections are easily created for children but after about the age of 11 new brain connections are impossible. We may use only what we have got.

From this point of view mental training in "direct vision" is helpful for children, as it allows their brain to develop, be more organized, and operate more effectively and faster. That is why children trained in this modality improve in their school lessons and have better attention to their tasks.

The next level of the informational exchange between the human being and the environment may happen without involvement of sensory perception. This mechanism is different from "direct vision". It implies interconnection of a particular brain with informational structures of the Universe. The explanation may be found in the principles of non-local interactions of quantum mechanics or Multi-Space Informational Entanglement[34] - different theoretical modeling for a New Scientific Paradigm. The Paradigm which will include Consciousness as part of the Universal Structure.

Fig.52. Direct vision.

ENERGY OF BEAUTY

One Fall, near the end of the 1990s, we had a big conference in Saint Petersburg dedicated to perspectives on science in the coming 21st century. I was one of the presenters and my ideas about light - photons - being the means for information exchange in living systems were very well received by people at the meeting. A part of my speech was about different types of structured water and their electrophotonic signatures. During a coffee break, a gentleman came to me and introduced himself as George Cioca, from the USA. He was speaking fluent Russian, but with a strong accent. I later found that he was Rumanian, settled in the USA and speaking six different languages. After discussion about water properties he asked me, would it be possible for us to make a blind test of different samples of water. I answered: "No problem. Please, send us your samples," and we exchanged business cards. When I looked at his card I was surprised to see that he is a Vice-President for research of the Estee Lauder companies in New York.

I was not thinking much about this meeting, but after a month we received a parcel with 10 little vials wrapped in aluminum foil. In every vial was water, and all vials were labeled from 1 to 10.

Prior to that time we had made some experiments with water testing, but in reality we had no thorough experience. So my claims at the conference were more like predictions than analysis of real results. Of course, we had some experience with microbiological cultures several years before. We studied spectral parameters of stimulated glow from yeasts in the process of cultivation. Results were very interesting, but it was just at the beginning of the collapse of the Soviet Union, no one was interested to proceed with this line of work, and it was closed, resulting in several articles and a Ph.D. dissertation. So the Estee Lauder water was a good chance to prove my expectations.

We divided the samples into two sets, and decided to make two independent lines of analysis in two laboratories using our equipment. When I captured EPC images from the samples, it became clear that we had two sets of different water. Bottles N 1, 3, 4, 6, and 9 contained normal tap water, and in bottles N 2, 5, 7, 8 and 10 the water was activated. I was very excited by this result and repeated every measurement 10 times to be sure. When I looked at the clock, I found that I had been doing these experiments all night long.

My friend - Alexander Kuznctsov - got very similar results, so we were very satisfied when we called each other in the morning.

I sent e-mail with our conclusions to NY and returned to everyday activity, earning a living. But several weeks later I received e-mail from Dr. Cioca inviting me to visit the Estee Lauder laboratories.

From that time on, I was offered a position as scientific consultant to the Estee Lauder companies. George Cioca's idea was to test possible applications of innovative technologies and ideas for the cosmetic industry, and I was offering them different ideas, crazy enough to attract George's attention. All our results were verified in NY by Dr. Glen Rein, a well-known expert in frontier science. Each year I visited Estee.Lauder laboratories with reports and presentations.

During one such meeting George introduced me to a gentleman: "Konstantin, this is Alex Vainshelboim, senior researcher of AVEDA company, one of our affiliates. He is your compatriot from Russia, so you may find joint topics for discussion."

Alex had a difficult life journey, common for a lot of Russian Jews. Being a young talented scientist in Moscow, he was unable to find a good position. In the Soviet Union Jewish people were not allowed to work in state research organizations. Officially everyone was equal, in reality it was a state policy of anti-Semitism. A young boy with a distinct Jewish appearance had no chance to get a good job in Moscow. After several years of ineffective attempts he decided to emigrate to Israel, together with his young wife, Tatiana. It took more than a year to get permission and at the border Russian custom officers striped them of all valuable things, under accusations of a smuggling attempt. They came to Israel without money, without jobs, with poor English. But they had friends and for the first three months they lived in a kitchen in a one-bedroom apartment of their friends in Tel-Aviv. Later came military service, a first job as a chemist, doctoral dissertation, their first child, and - finally - a job offer from the US cosmetic company. In the year 2000 Alex was a senior researcher at the AVEDA Company.

From the beginning of our work together, Alex was very suspicious and skeptical - this is typical for a real Jew. After a year of joint experiments he was able to persuade himself that electrophotonics is really a powerful tool. We became good friends, traveled together to different parts of the world presenting our findings to the International scientific community at different conferences and congresses, and in five yeas had published about twenty-five papers. This was all made possible only through the understanding and support of Peter Matravers, Vice-President for Research and personal attention by AVEDA President Dominique Conseil. And, of course, due to the professional work of the yang AVEDA's researchers, first of all Kent Momoh, Michael Hayes, Corissa Raats and Kathy Price.

What were the main results and ideas discovered in these years?

For Millennia, people have used cosmetics in all human societies. Wild tribal people painted their faces and bodies, sometimes even teeth. Ancient Egyptians invented perfumes and creams, and these goods were always the most precious commodities in international trade. In the Twentieth century, cosmetics became a huge multi-billion dollar industry. But it was always the idea that cosmetics are mostly for skin-care.

AVEDA proposed an idea that cosmetics influence not only skin, but also the person as a whole, mostly if applied with a massage in a combination of different procedures.

Skin is not just a cover, keeping our internal organs together, but is also an organ by itself, designed to exchange chemicals, energy and information with the environment.

We have billions of nerve sensors on our skin, and these sensors react to temperature, touch, and light. (You may recall our story about direct vision - blind children who are trained to see with their skin). And all the chemicals that we have on our skin somehow penetrate into the body. The condition of the skin is an indication of health, and all cosmetics applied to the skin have direct impact on the peoples' conditions. We decided to prove this idea in a classical way.

A group of volunteers - young students - agreed to participate in the experiment. They were divided at random into two groups - experimental and control. Each person was given a jar of cream and they agreed to apply it to face and neck every day in the morning and at night. The cream was expensive-looking and had a good smell, so people accepted this task with pleasure. We measured their physiological parameters and mood state before the experiment, and then a month after regular use of the creams. There was one difference between groups - the experimental group were using innovative AVEDA cream, and the control group were using neutral Vaseline. After processing all the data, it was found that parameters for people in the control group had either not changed, or in some cases had even decreased (it was winter time, and cold and darkness had depressive influences on people). Most participants in the experimental group showed statistically significant improvements in the measured parameters.

It was clear evidence that proper cosmetics had holistic effects on the person as a whole, with her or his physiological and psychological condition, i.e. both body and spirit.

It may be strange at first glance, but understandable if we accept the idea of informational exchange using skin. Application of a good cream helps eliminate contaminations from the skin, protects from dust and sends positive signals from the cream ingredients. The process of cream application is itself very important. If it is done with smooth, gentle touching, it sends additional calming information to the sensors, creating signals of peace, calmness and relaxation. The body receives these signals, accepts them as pleasant, positive information, and transforms to a relaxed, calm state.

So it was statistically demonstrated that cosmetics have holistic effects for the body. Of course, it should be good cosmetics. And usually we associate this quality with the idea of natural products, created from the flowers, plants and grass, but not from petroleum. This is our intuitive feeling, but perhaps it is just our imagination? Maybe there are really no differences between natural and synthetic stuff. If we like the smell and the color, what is the difference - whether it originates from the Amazonian forest or from a Chicago chemical laboratory?

Studying the oils

We decided to test some of the key cosmetic ingredients - essential oils. For centuries oils have been extracted from natural stuff and these oils became an extensive science and industry. In the 20th century chemists became skilled enough to decode the composition of these natural substances and mimic nature by creating chains of polymer molecules similar to natural compositions. In some cases it was tremendously successful - as in the case of rubber. This invention saved million of acres of Amazonian forests, as it allowed manufacturing car tires chemically without using tree tar. But applications to food and cosmetics - that is a whole different story. So, are there some differences between natural and synthetic products having the same chemical composition, or are they absolutely similar, so we do not need to spend natural resources to produce expensive ingredients? This was the question of our research.

Oils used in personal care products can be obtained by either of two conventional ways. They can be synthesized, by using precise organic chemical synthesis techniques. On the other hand, the oils can be naturally derived, predominantly from plant sources. In either case, innovative technologies are being applied in order to get the purest ingredient that is necessary for obtaining the desired oil. These techniques use conventional chemical and physical analytical tools that by themselves cannot distinguish between the same oils of natural and synthetic origin.

The problem of detecting individual differences of chemically similar liquids, as well as highly dilute aqueous electrolyte solutions, continues to be unsolved for various areas of the natural sciences, such as medicine, biology, nutrition, and cosmetics.[35] For example, some certain subtle differences of smell and taste are very difficult to detect by using conventional methods of analysis, such as gas chromatography (GC) and the like (except under unique advanced techniques that are expensive, time consuming, and are not readily available). However, a trained human nose or mouth of a perfumer or taster can detect differences of that kind.

The technique of assessing liquids by investigating characteristics of the gas discharge around drops of those liquids had been shown in previous work.[36,37] These studies demonstrated that strong electrolyte solutions, such as NaCl, KCl, $NaNO_3$, and KNO_3, have differences in the characteristics of stimulated light around the drops of liquids. These differences were found both in comparing differences between neighboring concentrations of one electrolyte solution and between the same concentrations of various electrolyte solutions. Controlled blinded randomized assessment of four split samples of 30c potencies of three homeopathic remedies from different kingdoms demonstrated that EPC/GDV technology allowed differentiation of the remedies from the solvent controls.[38]

We selected several compositions of essential oils, having the same mass spectrum, but being natural or synthetic origin. Then we had to develop a way to test them in our laboratory.

We had developed technology to study photon emission from water, and it was quite effective in selecting between different types of natural, tap, or processed water. The topic of structured water became very popular recently, mostly after beautiful pictures, created by the Japanese researcher Yamoto. But in our water testing technique we kept water in a standard syringe, while for precious essential oils the tested quantity would need to be as little as possible.

After several attempts we developed an approach where a drop of test solution was placed in a little hollow of a glass plate and a thin wire touching the liquid served as an electrode. The glass plate was positioned on the optical lens of the EPC instrument for measurement. The main idea was to measure the time dynamics of the stimulated photon emission during several seconds. This idea was based on fundamental principles of life.

We live in a time-space domain. This means that time is an active power of our life process. Time has particular meaning for biological subjects: our inner time is different from astronomical time. Remember, how slow was the time in childhood, how long were periods from one birthday to another. And how short they have become when you have grown. To slow down inner time - to prolong young and mature years - this is a long-lasting dream of humankind. All biological processes have time development and new features may unexpectedly evolve through time. Such features may appear if you apply some dynamic testing and monitor the behavior of a subject in response to the test. For example, in sport medicine it is obligatory to test athletes before and after physical loading and monitor their reaction. EPC, in principle, is a dynamic test, where we monitor response of a subject to the applied electrical field.

Developing a testing technique for essential oils took us a long time, to get stable reproducible results. But we finally succeeded in producing highly reproducible beautiful images of different oils. Then we began testing different compositions.

From the first glance EPC images of natural and synthetic essential oils looked the same and in most cases had the same parameters: area, brightness, entropy. But when we tested them for longer periods of time - 10, 20, 30 seconds - their parameters started to differentiate. An impression was that after a relatively long period of testing the intrinsic, hidden features had been revealed by the natural substances. When we discovered this for the first time, we had, perhaps, the same feelings as an art work restorer, when cleansing a simple painting he sees a glimpse of a layer of older, much more valuable colors underneath a modern picture.

Please, look at the picture. Figure 53 shows a graph of the averaged intensity parameter for Oil of Bitter Almond and Synthetic Benzaldehyde. Fisher's Test was used at 95% Confidence Intervals for every moment of registration of the dynamic processes for the two liquids. The graph clearly demonstrates that the average values of the realizations for Oil of Bitter Almond and Synthetic Benzaldehyde show a statistically significant difference for the averaged intensity parameter after four seconds of measurements.

Another example. The oils of Moroccan Rose, Bulgarian Rose, and Russian Rose are chemically similar. They are, however, sourced from different locations. During the initial moments of measurement, the time series for area for each oil coincide with each other. However, in less than one second, each of the curves began to deviate and differ statistically significantly from one another (see figure 54). Parameter "a" in the approximation for the different Rose oils was $2*10^{-2}$, $7*10^{-3}$ and $-2*10^{-2}$, respectively. Positive and negative values of "a" for Moroccan Rose oil and Russian Rose oil, respectively, with the same absolute values of area illustrated the sensitivity of this parameter to the physical and chemical properties of the liquids. In addition, no statistically significant difference between Moroccan Rose and Bulgarian Rose oils was found, but they both differed statistically significantly from Russian Rose oil.

Several different combinations of oils have been studied in this research line:

- Oils of Essential and Synthetic Nature.
- Oils of Organic and Regular Origin.
- Oils Received in Different Climatic Conditions and Extracted by Different Ways.
- Natural Oils: Fresh and Oxidized in Various Ways.

And in most cases statistically significant differences were found between samples.

These results demonstrated that in complex liquid substances not only chemical composition, but also structure of a liquid is essential as well. In the process of chemical synthesis scientists may precisely replicate the chemical composition of a substance, but some details of the natural stuff are still unique and unpredictable. We can feel the difference between wine from the same grapes, produced in different years and in different regions of the world, the same as we can detect difference in natural and synthetic cosmetic ingredients. And I am absolutely sure that in the near future strict scientific research will demonstrate that using natural products is extremely advantageous for our health.

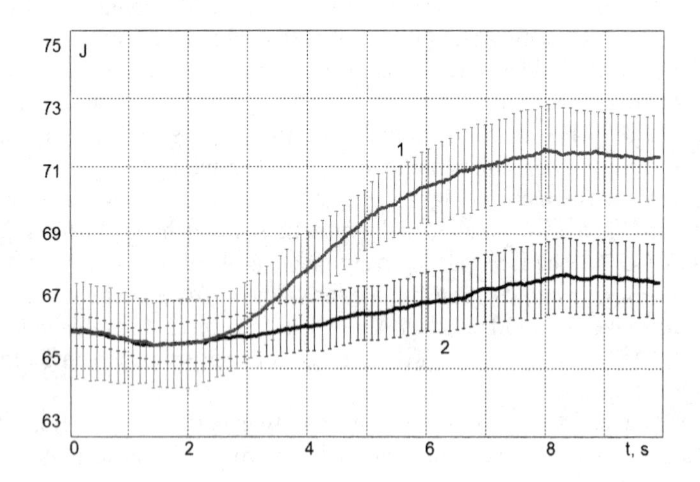

Fig. 53. Time dependence of the intensity of the gas discharge image around drops of Bitter Almond (1) and Synthetic Benzaldehyde (2) oils.
Data averaged on 10 measurements.

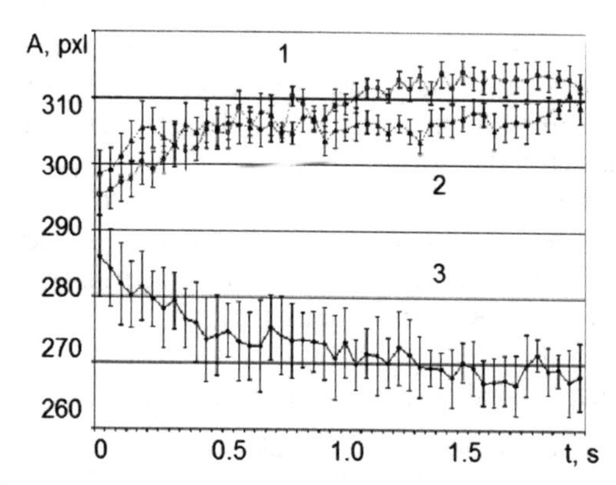

Fig. 54. Time dependence of the area of gas discharge image around drops of Rose oil of different origin. Data averaged on 10 measurements. 1 - Morocco Rose; 2 - Bulgarian Rose; 3 - Russian Rose.

Studying the Gemstones

Gemstones have a long history of use in healing. Although they are mineral in composition, their highly ordered structure, visual appeal and unique effects on visible light have led many to experiment with their effects on the human body.

In another language, the gemstones' highly ordered composition allows them to conduct and align subtle electromagnetic fields. This has had two important holistic applications: structuring homeopathic liquid materials and aligning bodily fields by direct contact with the skin.

The latter technique is quite old. Gemstones find one of their earliest uses in the Ayurvedic system of India. Ayurvedic texts describe the use of gemstones to balance chakra points on the body, correcting energetical imbalances believed to cause disease and discomfort. Tourmaline has been held in Ayurvedic tradition to act upon all three doshas, or personality types, a rare trait. Gemstones have been used in the tradition of Ayurvedic medicine for centuries.

Ayurveda was already a mature system by the time it was discussed at the councils held after the Buddha Gautama's death c. 540 B.C.E. Indian systems of medicine were later recorded and called tikiccha by Sri Lankan scribes recounting the council. Despite the different name, this is clearly the same system, with 11 of the 12 branches considered to be the defining applications of Ayurveda preserved intact.

The first involved discussions of the use of gems in Indian healing appear to have taken place in the 6th century C.E. The cornerstone text Brhatsamhita, by Varahamihira, was composed c. 530 - 580 C.E. A second text, more directly focused on the use of gems, was the Ratnapariksa by Buddhabhatta. This text's date of creation is under dispute - many scholars date it as early as the end of the fifth century or the beginning of the sixth. However, some have argued that it could have been written as late as the 12th century.

What is not under dispute is that in the early 1300's, Candesvara wrote the Ratnadipika, the most focused and authoritative discussion of the use of gems to that point. While it builds from the material found in the Brhatsamhita, it expands it and represents the basis of gem selection to the present day.

According to Ayurvedic belief, gemstones act to balance or tune specific Chakras. In AVEDA laboratories gemstones were selected and prepared according to Ayurvedic tradition. We consulted with Ayurvedic practitioners in the Himalayan region, observing many of these preparations firsthand.

Table : Claims and preparations for selected gemstones

Stone	Herbs used in preparation	Ayruvedic Claims
Quartz	• Keora (Pandanus odoratissimus) • Himalayan water	• Universal conduit • Amplifies, focuses, stores, transforms, energies. • Stimulates psychic perceptions • Attunes to all chakras, all signs and all numbers
Topaz	• Rice (Oryza sative) • Horse gram (Dolichos biflorus) • Himalayan water	• Stone of the sun • Assists with the elimination of toxins • 3rd chakra and higher
Tourmaline	• Rice (Oryza sative) • Horse gram (Dolichos biflorus) • Himalayan water	• Brings healing power to the user • Balances left and right brain • Gives discernment and "sight" into a given situation • Attunes to chakras of like color
Ruby	• Lemon (Citrus medica) • Himalayan water	• "Queen" of all gemstones • Stone of love • Helps heal the heart and all blood impurities • De-toxes the body • 4th chakra
Emerald - Grade I	• Himalayan water	• Healing of eyesight and speech impediments • Tranquilizing effect • 4th chakra and higher

Because Ayurvedic claims were the basis for the selection of the gemstones, gems were obtained from the Himalayan region, and Ayurvedic doctrines dictating the preparation of the gemstones were followed closely. These preparations involve washing, boiling with herbs, and rinsing. In all stages, water, herbs and materials required for the process were obtained from sources in the Himalayan region from which the gemstones were obtained.

Specifically, the story of gemstones deals with a tradition that has been used by many cultures for over 5,000 years. For example, the Ayurvedic philosophy, which is based on ancient writings from Sanskrit, includes great details about this story of gemstones-that is, how, why, and in what ways they are being used for daily living. The story deals with the concept of gemstone therapy, which basically describes gemstones, in relation to energy, as a means for stimulating healing and realignment within one's body. Such crystalline stones have the ability (and energy) to amplify areas of energy that are already present in the body. To be specific, gemstones have been used to tune and amplify the electromagnetic energy that flows naturally through the organs in one's body. Different gemstones are more energizing for different organs and/ or meridians; however, some are beneficial for many areas of the body. In short, gemstones are also known to improve the balance, health, and general well being of one's body.

Gemstone elixirs are one application of gemstones described in these cultures. Elixirs are generally aqueous solutions that consist of various herbs, organic citrus juices, and often times, gemstones. Ayurvedic medicinal practices include the preparation and use of gemstone elixirs. Strict Ayurvedic procedures for preparing these elixirs actually came from original Sanskrit writings that were discovered thousands of years ago.

The rationale behind the principle of gemstone elixirs is quite simple. The energy that is stored in a gemstone can be transferred to a fluid medium such as water. Water has "good memory;" that is, it absorbs and maintains the characteristics of any object that it contacts. Therefore, in the case of gemstones and in preparing the corresponding elixirs, water will absorb and maintain the energy of any gemstone that it contacts without "draining" the gemstone of its own energy. As a result, the structural properties and characteristics of the water change in such a way that its energy would become amplified.

Basically, a gemstone elixir thus prepared, can be used for the spiritual healing of one's body. The energy of the gemstone (not the gemstone itself) that now lies active within the water can be passed from the elixir to the body. Again, without draining the elixir of its own energy, the body would receive and maintain this energy, and thus its own energy would become amplified in such a way that it would provide spiritual healing to the body.

Therefore, based on this information, it would make sense to use a tool like Electrophotonics in order to evaluate the energy levels that exist within gemstones. Going even further with this concept, such a tool could also be used to evaluate the performance levels (in terms of energy, of course) of the corresponding applications of gemstones. Such applications include those of gemstone elixirs, gemstones that are ground into fine powders (gemstone powders), and the like.

Various gemstones have been evaluated for their respective energies. Images at fig. 55 pertains to that of the gemstone of Golden Citrine. To the left is the actual gemstone of Golden Citrine, and to the right is the energy given off by the same gemstone.

Next set of images pertains to that of the gemstone of Pink Tourmaline (see Fig. 56). To the left is the actual gemstone of Pink Tourmaline, and to the right is the energy given off by the same gemstone.

Now, based on the already given results, one should understand that gemstones do in fact have energy, and that Kirlian Photography has allowed one to observe that. However, what about the gemstone applications; can this tool be used to evaluate the energy or performance levels of these various applications? The answer is yes; one such gemstone application is that of gemstone powders, or gemstones that are grounded into fine powders. In a powdered form, the gemstone is actually still composed of its same material; however, it differs in its physical appearance. Regardless, the energetical properties of the powdered gemstone should still remain intact.

Another gemstone application is that of gemstone elixirs. Like gemstones, various kinds of gemstone elixirs have been evaluated for their respective energies in comparison to that of USP Purified Reverse Osmosis (RO) water. The next images that will be presented in the following demonstrate that this method does allow one to observe an important phenomenon about gemstones (see Fig. 57). That is, gemstones do in fact add energy to the substance or substances in which the gemstones are being applied to. Keep in mind that these gemstone elixirs were prepared from their corresponding gemstone powders in RO water following strict Ayurvedic procedures taken from original Sanskrit writings. What is being observed that each of the gemstone elixirs gives off more energy than that of the control RO water.

A final gemstone application that will be discussed is that of incorporating a gemstone powder into an already existing base product, such as a skin cream or lotion. Analogous to the studies that have been done on gemstone elixirs, various kinds of gemstone powder applied products have been evaluated for their respective energies in comparison to that of the base product that does not contain any gemstone powder or application with Tesla Coil Kirlian Photography. The next image that will be presented shows the energy comparison between a skin cream charged with Pink Tourmaline powder and the same skin cream that does not contain any Pink Tourmaline powder or any gemstone powder for that matter (see Fig. 58). Once again, gemstones do in fact add energy to the substance or substances in which the gemstones are being applied to.

Therefore, if all of the previously mentioned data are taken into consideration, then it can be quantitatively shown how these five gemstones compare to each other in terms of glow area pixels, and thus energy. Ranking the five studied gemstones in the order of highest glow area pixels to lowest glow area pixels, Yellow Topaz is first, Garnet is second, Aquamarine is third, Golden Citrine is fourth, and Amethyst is fifth or last. Again, one can then assume this same order of rank in terms of energy for the same five gemstones. Therefore, Yellow Topaz has the highest energy, followed by Garnet, Aquamarine, Golden Citrine, and Amethyst which would then have the lowest energy.

As we see from the above material, gemstones have significant effect on all media, where they may be immersed, be it water, oils, or creams. Then it is easily understood, that application of gemstones to human body should have an effect as well. And this effect was really measured and statistically verified. But is it really surprising? Humankind has known this for centuries, and jeweler's art has always been and for a long time will remain an important part of our culture.

A very important role in the effect of any cosmetic product belongs to smell. We may be suddenly attracted by a faint smell of perfume at a crowded dinner-party and a memory of a long-forgotten event flashes before our eyes. Smell has strong effect on

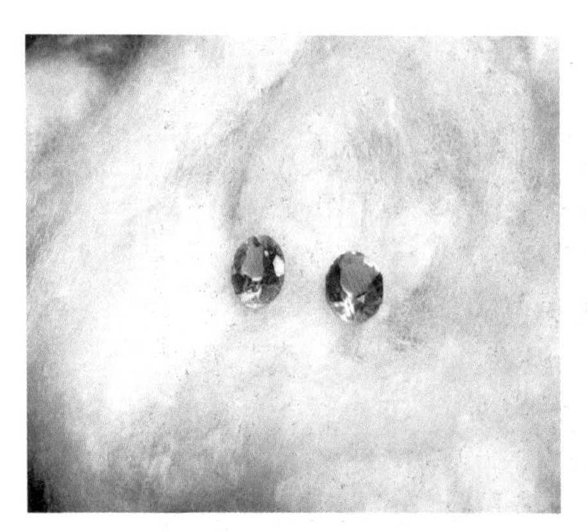

Fig.55. The actual gemstones of Golden Citrine are shown on the left, and shown on the right is the energy given off by the same gemstone (Golden Citrine) as a result of Tesla Coil Kirlian Photography, thus showing that Golden Citrine, a gemstone, has energy.

Fig.56. The actual gemstone of Pink Tourmaline is shown on the left, and shown on the right is the energy given off by the same gemstone (Pink Tourmaline) as a result of Tesla Coil Kirlian Photography, thus showing that Pink Tourmaline, a gemstone, has

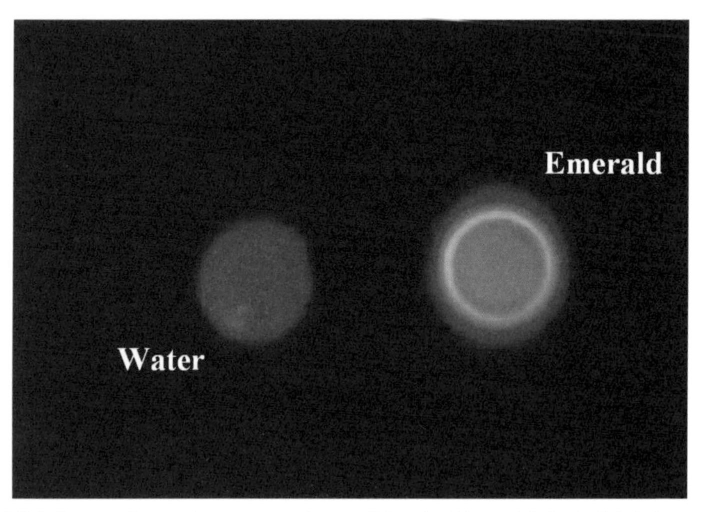

Fig.57. This image shows the energy given off by the Emerald elixir (right) compared to the energy given off by the control-USP Purified Reverse Osmosis (RO) water (left), both as a result of Tesla Coil Kirlian Photography. This image definitely shows that the Emerald elixir, as a gemstone elixir, gives off more energy than that of the control RO water.

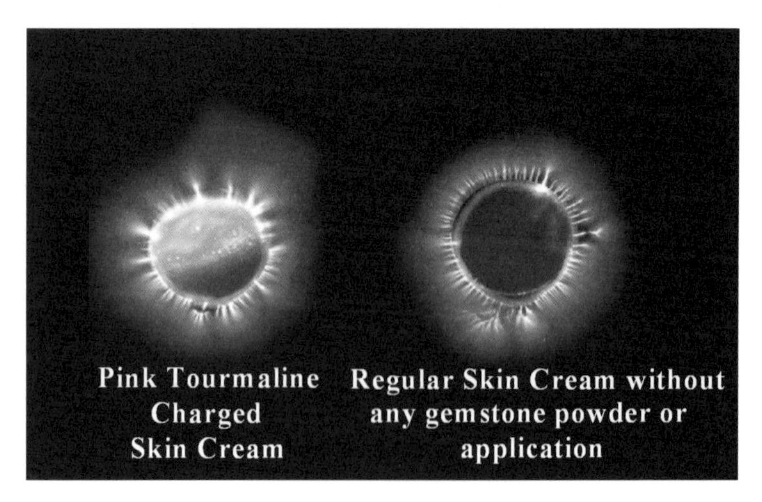

Fig.58. This image shows the energy given off by a skin cream charged with Pink Tourmaline powder (left) compared to the energy given off by the control-the same skin cream that was not charged with any gemstone

people, at the same time we clearly feel smell at a moment's notice but lose this sensation a moment later. Why is it so? What is the role of smell in communication of a person with the world?

Study of the olfaction process is very complicated, and no wonder that quite recently a smell study received the Nobel Prize award. We decided to see whether we can detect the reaction of Human Energy Field to smell. Big series of experiments demonstrated that we can detect reactions to smell of different essential oils; reactions correlate with preferences and - what is most interesting - they are different for natural and synthetic oils. It was found that intensity of reactions correlates with psychological, emotional and physical characteristics of a person, that our results reflect classic descriptions of physiology of smell - adaptation: a period of initial excitation followed by a declining activity represents a decreasing sensitivity to aroma. The adaptation process takes longer for the aromas considered unpleasant. Objectionable odors cause GDV-measured reactions longer than pleasant aromas. Objectionable odors show greater difference in the results between samples.

All these results are of great interest, as they support the concept that reaction to smell strongly depends on the psycho-emotional condition of a person. That means that in different emotional states people may have different smell preferences. On the other hand, response to odorants may strongly depend on the overall comprehension of situation (i.e. environmental conditions, type of perfume bottle, package, etc.). The more we understand the inner mechanisms of smell reactions, the more attractive cosmetics and perfumes we will be able to create.

Energy of Human Hair

Hair provides no vital function for humans, but its psychological effect is nearly immeasurable. Luxurious scalp hair expresses femininity for women and masculinity for men. The lack of scalp hair or the presence of excessive facial or body hair is often as distressing to females as the loss of beard and body hair to males. Study of hair and its response to various treatments and stimuli is of tremendous importance for the humankind.

Each type of hair undergoes repeated cycles of active growth and rest. The relative duration of each cycle varies with the age of the individual and the region of the body where the hair grows. At the same time hair has traditionally been considered to be an inert tissue. While the hair root is vital, its product, the hair shaft, is primarily dead keratinized cells. While the human hair can be colored, styled and cut to a dramatic effect, its direct interaction with the body has been commonly believed to be nonexistent.

The initial experiments tested hair on live models, along with testing swatches single-sourced from their hair. The results of these GDV tests showed significant differences between hair in vivo and in vitro.

Live models consisted of adults, both males and females, aged 18-55. They represented a variety of hair types; with damage levels ranging from virgin hair, to hair that had been treated or styled extensively. The test subjects sat near the camera so that their hair could pass completely across the electrode. The initial ten-second reading was taken with their hair on the electrode. A few minutes later the hair was cut between the electrode and the subject, detaching the hair from the subject. The cut was made in a manner to ensure that the sample did not move on the electrode (see figure 59). In many cases a distinct reaction of hair photon output to the cut was registered (see fig. 60).

After the initial reading, 10-second time series were measured at specific intervals. Readings were taken at? with the intervals of?24, 98, 121, and 165 hours. The averaged readings from each series were plotted together to show the change in the intensity reading over time. See figure 61.

Over 100 swatches, mannequin heads and other check samples were used as hair references. Swatches were single-sourced, assembled by International Hair Importers and Products, New York, USA from hair that was collected during the tests. The mannequin heads, both single-source and multiple-source, were obtained from Pivot Point International Incorporated, Illinois USA. Numerous experiments, conducted with samples of this dry hair demonstrated high stability of their photon output in time. It means that dynamic reactions were registered only for fresh live hair.

Similar results were found for time dynamics of laser light reflection by hair: different behavioral pattern of hair optical parameters for freshly-cut and dry hair was found.

These results demonstrated an important fact: hair is a biologically active matter, having reactions to the environmental influences. We may speculate that **hair may serve as a way of communication between human and environment**. Hairs react to the environmental influences and send signals to the body. Then it is understandable that if a signal existed between human hair and the body this signal would be expected to weaken after interruption of this connection.

Why do we mostly have hair on three parts of the body? Because these parts are important for receiving outside signals concerning subconsciousness information to the head, the signals related to sex and to the sense of danger.

Why do women typically have longer hair than men? Because women are much more intuitively sensitive to the environment and to other people, they "feel" good and bad people around and have different attitude without any logical reason.

On the other hand, men of which type have long hair? Typically these are people of art: artists, singers, painters - these are people who need response from other people. While businessmen, scientists, athletes typically have much less hair, and quite often demonstrate bald scalp - they are concerned about their own brain-work and rely on their own decisions, often regardless of the others' opinion.

Of course, this is just a hypothesis, which needs a lot of attention and further proofs, but it opens up a whole new perspective in studying and influencing **the "Life Force" of hair.**

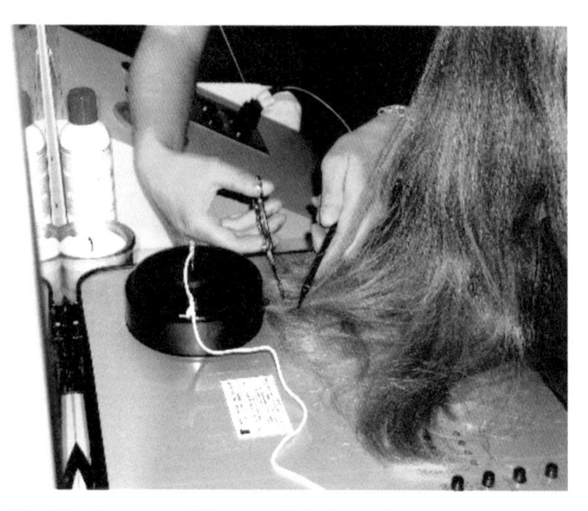

Fig.59. Cutting of live model's hair during GDV test.
Electrode is under circular black cover.

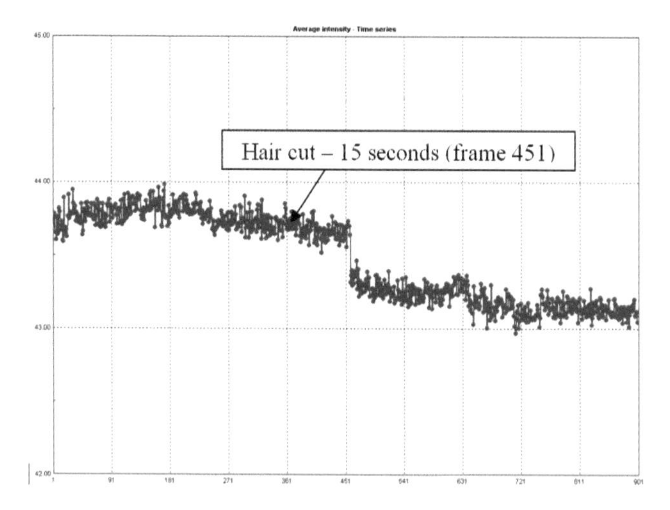

Fig.60. Example of a 30-second reading, with hair cut from the model midway through the reading.

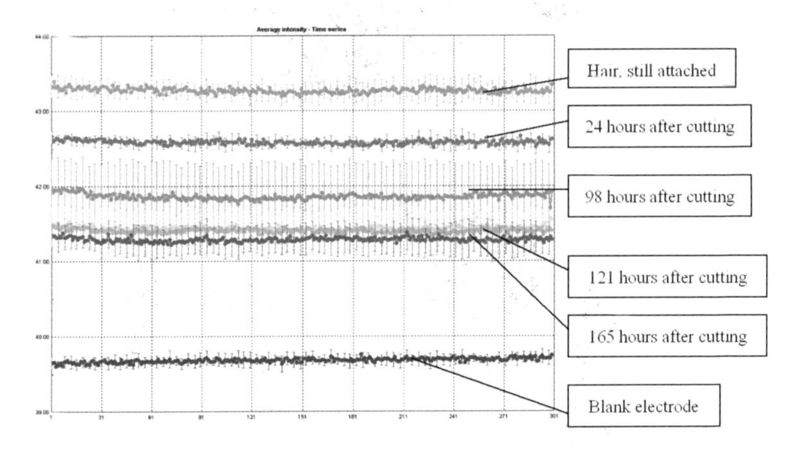

Fig.61. Intensity readings from a live model, showing typical trend. Note gradual and consistent decrease in GDV intensity over this multiple-day test.

Part III

THE LOST WORLD

IN SEARCH OF THE ENERGIES OF LIFE

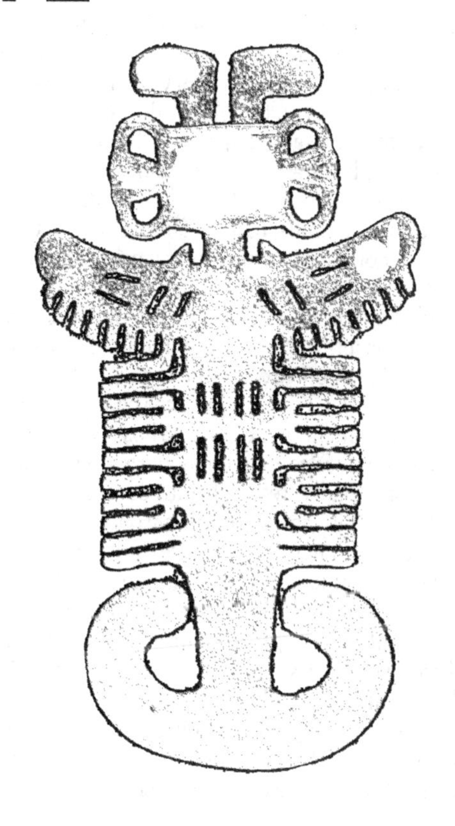

WHY DID WE GO THERE?

For many years now, the main theme of our research has been the study of human energy and the world around. During research, we inevitably came across an understanding of geo-active zones, and people who were studying them. I won't get bogged down in a detailed process of this problem, so as not to digress from my main theme. It is important to note that apart from the geopathic zones which journalists are so fond of, there are beneficial geozones, which enhance the mood, feelings and emotions of the happy visitor. It is these zones which are of interest to us, not just because you can get high for a while in the lap of nature, but most importantly because therein lies one of the golden keys to the treasure trove of health and longevity. This problem has exercised humanity throughout history, and even now we are a far cry from a model of a healthy lifestyle.

Do geoactive zones really influence human energy?
To what extent does this influence depend on environmental factors?
How do the man-made elements of our civilization influence this process?

Research into these and a whole range of other issues represented the challenge of our expedition. A separate challenge was researching water, the magic liquid, the foundation of life, health and biological diversity of our Earth.

There are many geoactive zones on the planet. Russia is rich in them like no other. Any rural church, monastery, or sacred spring is linked to geoactive zones. Many times we travelled to such places and took measurements. The latest major event took place on the Solovetsky islands. But another vitally important factor drew us to the Himalaya and Roraima regions, namely the absence of technogenic influences. No radiophones (mobile communication only works within cities in Nepal and Venezuela), no FM radio stations, no television. And, of course, there are no factories or industrial plants and their constant emissions in our lungs. Such places can only be found in the remotest corners of the earth, which are not so easy to get to.

It goes without saying that there are not the throngs of people in these areas as to be found in our 'civilized' regions. For people have their field too, and by all accounts, the collective field must exert its influence on all those who find themselves within it. It is an extremely interesting question which modern science has not yet touched on, so there is work to be done. Sadly, research into such issues currently remains the reserve of enthusiasts. But happily, there are still untouched corners of untamed nature on earth, which are out of the range of collective and radio fields. And from time to time we set off for such places. Among the first was our expedition to Nepal, to the foot of Everest. Next year: the Solovetsky Islands, those secretive northern labyrinths. And finally Latin America, a continent of great civilizations, of proud but friendly people and virgin, unbridled nature, with its jungles, savannahs, thousand-metre-high mountain peaks, often with a distance equivalent to an hour's flight between them. We were lucky though: there remained many corners on earth untouched by civilization. To research them would take more than one lifetime, and the results are of interest to many. It is to those people that these expedition memoirs are addressed.

THE HIMALAYAS

"The Hiamalayas, "Home of the Snows", the most impressive system of mountains on the planet, and for centuries the setting for epic feats of exploration and mountain climbing, are a world unto themselves, the home of ancient and mysterious cultures, where time and space have meaning unlike anything that we have come to know elsewhere.
This is a land where spiritual presences hover and can be felt, where magic and reality, history and legend, are indistinguishable. There are lost valleys where genuine Shangri-Las can still be found, where one can take refuge and let one's life flow along, calm and tranquil. Here religious elements are common features in the landscape; mountains and nature herself are simply altars to the gods, and the personification of deities."
Marco Majrani. Himalayas.1994.

All the surrounding space is filled with radio waves. Television, radio, mobile phones, technical systems - all that creates a permanent background where we live, sleep, and work. It remains only to hope that it does not influence our human gene pool greatly. But in order to organize precise experiments we sometimes wish to have conditions free from this electromagnetic pollution. Therefore, after long discussions, we decided to go to the High Planes, to the Thin Air areas. There are neither mobile phones, nor television there. In addition, we were interested in a number of professional questions: At what altitudes can EPC measurements be done? How will the change of air pressure and number of charged particles influence EPC-grams with gaining height? Are there natural limits for carrying out EPC measurements? And, finally, what changes of state can be fixed at high altitudes?

In order to answer all these questions, and also to perform a series of experiments, a small reconnaissance expedition to the Himalayas, Nepal was organized in May, 2001. The region of Everest was chosen, as heights of 5000 m or 17000 ft and more can be reached in that very region, gradually gaining the altitude, which provides step-by-step acclimatization. The fact is that the rarefied air in the mountains causes altitude sickness attacks, manifested in headaches, weakness, and in loosing one's consciousness, at the worst. So, if an urban person is taken to the height of 5000 meters by a helicopter, he would most probably loose his consciousness. At the same time, if one climbs this altitude during a week, the organism has time to get used to the mountains and stands the lack of oxygen quite well. Although then, having come down, it reacts with an abrupt increase of immunity and energy. Mountain trainings are used for the preparation of Olympic teams.

Our group consisted of 9 people: 7 men aged from 32 to 57, and 2 women aged 21 and 23. 3 men had had previous experience of climbing peaks higher than 6000 m, the rest had had no mountain training. We used 01M EPC camera model fed from 12 V battery and "Toshiba" laptop for the experiments and registration of EPC-grams.

From Katmandu (the capital of Nepal) to Lukla (2840 m or 9300 ft), unreachable by motor roads, we took a small airplane and then a helicopter. Coming from Lukla uphill you can find fewer and fewer traces of our civilization. One can continue the way only on foot from there, by mountain tracks, gaining height with every day. To our delight, on the first day the track ran along a flat river valley, somewhere even downwards. The sun, bright in the beginning, changed into a downpour, but that did not prevent us from reaching the village Phakding (2640 m or 8660 ft) where the first EPC measurements were done. Everyone was yet in a good state of health.

It is worth mentioning that the possibility to use sherpas, local porters, is a big advantage for trekking in the Himalayas. Carrying loads is the only source of income for them; therefore, the Nepalese government stimulates this occupation in every possible way and creates the most favorable conditions. Tourists from many different countries slowly make their way along the rocky paths accompanied by the ever-cheerful sherpas.

The next day a lengthy rise to the village Namche-Bazar - the capital of the Himalayan sherpas, lay ahead. We had to climb more than 800 m in a few hours and reach the mountain settlement situated at 3440 m or 11280 ft. These are quite high altitudes, and one of our participants got evident indications of mountain sickness: headache, dizziness, lack of energy, and absence of appetite. Walking at such altitudes, even without a rucksack, is a serious physical load, especially for a person used to spending most of the time in an armchair, and not in a gym. So, we had time to rest and to have a snack when the last participants appeared.

The Sagar-Matha National park is an ideal place for traveling. One can come here with a sleeping-bag and a spare T-shirt. Mountain cabins, where you are offered a room and more or less acceptable food for a cheap price, are all around. Although, there is no heating in the rooms (and at night the stem of the thermometer often falls below zero), and you wouldn't like to look at the local food after three days, but all that is a mere trifle compared to the beauty of the mountain landscapes. At night, by the light of candles, we were doing EPC measurements; luckily we had opportunity to charge batteries in most of the villages.

A more difficult trial lay ahead next day: a climb to Tengboche monastery, to the height of 3860 m or 12660 ft. Imagine: on the top of a woody hill, surrounded by ice peaks, red structure of a Buddhist monastery hovers above mountain valleys. Several hours following the steep mountain tracks to the closest village, only monks live near the monastery. In the recent decade, civilization has given touches to the surrounding landscape: a few lodges (mountain hotels) for travelers from all over the world have been built near the monastery, and a satellite antenna has been set up the hill. However, the same as 1000 years ago, at 3 p.m. the traditional ceremony starts, during which monks in saffron frocks offer up prayers to the Buddhas, providing eternal renascence of the Universe. The monks murmur prayers, blow trumpets, beat drums, and repeat this procedure day by day - not for the parishioners, not for the visitors, but for the high Gods which are the only ones who hear the sacred texts. And Machu Picchu, the sacred city of the ancient Incas, situated at the peak of mountain range in the heart of the Peruvian Andes comes to mind. Purposes served by this hard-to-reach village, far from the main trade routes of ancient Inca are still being debated even now, but the example of the Himalayan monasteries shows that the main purpose of such places is serving the Gods, and it is better to be rather far off from human vanity to do that.

The steep, many-hour ascent to Tengboche monastery, to the height of 3860 m or 12660 ft went along picturesque canyons; the river was glittering in the sun and the blossoming rhododendron trees were standing on the sides. The load and altitude adversely affected the health of three more members of the team - they hardly managed to get up to the top of the monastery hill. Some of them might have been damning the day when they had made up their mind to go on the "Himalaya trip." Fortunately, the next day we didn't have to climb anywhere: we planned to spend it near the monastery, walking around and making measurements. (Fig. 62, 63)

One of the key moments there was the arrangement of sessions of telepathic communication with Moscow. We considered that the special monastery conditions would contribute to the success of communication. We had been absolutely correct, as it was found later in Russia, after the processing of experimental data.

The next day was spend in measurements, experiments, processing of previous measurements, and slow walks along beautiful landscapes around Tengboche monastery. At night our souls were blessed during a long picturesque Buddhist ceremony.

In the morning our team divided into two parts: the four went down, and the five, the healthiest, started the way up to the Everest base on foot or by helicopter. The day was very foggy and the helicopter was not able to bring us up. What we had to do was to climb up 3 hours with heavy rucksacks in the cold mist and snowstorm. There, at the altitude of 4910 m or 16100 ft, on the Khumbu glacier moraine, the Lobuche mountain hotel was situated. Views to the rocky glacier and the surrounding slopes opened from the windows of the small rooms, but the inhabitants of the hotel spent their free time in the cafe - the only place where the big metal furnace was being stoked and one could order a glass of hot Nepalese tea. We spent three days in Lobuche, climbing up to 5600 m or 18300 ft, gathering samples of water, and carrying out EPC measurements in the evening. Such altitudes, followed by insufficient acclimatization, inevitably tell upon physical condition: sleep becomes troubled, many people feel temple-ache, and physical efforts go very hard. But all this was nothing to the magnificent view of icy peaks, rising all around up to 8848 m or 29000 ft, where the ever-deep-in-clouds Everest, the sacred Chomolungma, stayed. After enjoying the exotics of the altitudes we started our 2-day way back.

Thus, we managed to make EPC measurements of the same individuals from the sea level up to 5000 m or 16400 ft, check the equipment and find out what was necessary for a more serious expedition. The main problem, as expected - computer powering. A laptop works no longer than three hours from battery, after which it has to be charged. As you understand, there was practically no electricity there, only sometimes we could charge computer batteries. The GDV Camera instrument was working only in battery mode and demonstrated quite stable operation. Now we can assert that the GDV Camera instrument can successfully be used at 5 000 m altitudes above the sea level and maybe higher.

So, what conclusions were we able to draw based on the results of this expedition? First of all, we made sure that it was possible to observe the change of state of the participants during the ascent and detect their reaction to the mountain conditions with the help of EPC bioelectrography. This result is not as trivial as it seems at first glance. The atmospheric pressure in the mountains is low; therefore it wasn't clear a priori whether that pressure drop would affect the parameters of gas discharge cascades used in the EPC technique. The calculations showed that evident effects from atmospheric changes shouldn't happen, because we extract the main information by stimulating photon and electron emission from the object itself. The measurements confirmed all these ideas. The EPC images demonstrated the change of state of the participants: they were significantly varying during the first days for those who suffered altitude sickness, and stayed practically the same for those who had had physical training. EPC parameters of all the participants showed significant changes at the altitudes close to 5000 m - the effects of insufficient acclimatization. The EPC image changes correlated with the characteristics of physical state typical of high altitudes: tumultuous dreams and shortness of breath when performing physical loads. Interesting that all the participants demonstrated an increase of energy levels in the region of the Tengboche monastery. It is yet only a single observation, we will need to determine if that is connected with the influence of the special place and the monastery itself, or that is just a psychological reaction. Unexpected data were received taking EPC images of the local people - sherpas. These measurements were only done for a few sherpas in the settlement of Lobuche at the altitude of 4910 m, as our time was limited. EPC images of all the measured sherpas were distinctly different from those of the Europeans (Fig. 64, 65)

In accordance with our classification 3 they belonged to IIb -IIc type, typical of the altered psychophysical state. What was that: their usual state or the state special for that very moment? Was that connected with certain physiological parameters, for example, blood pressure or special structure of skin? At the present moment, we don't have answers to these questions. The obtained data are insufficient for any kind of conclusions, but determine the directions of future research. Now it is clear what shall be taken in the next expeditions.

Near Tengboche monastery, we carried out several sessions of telepathic communication between Nepal and Moscow. We measured the operator with the help of EPC and EEG, and similar measurements were done in Moscow. The signal was sent to Moscow at an agreed time, and we were registering the state of the operator during signal transfer. As soon as we returned to Russia, all the material was analyzed and successful contact was shown in 8 cases out of 10. If research and practice continues in this way, people may start sending telepathic messages to one another and organize telepathic conferences by mid-century. Who knows?

Fig.62. GDV measurements in Nepal.

Fig.63. Tjangboche Monastery.

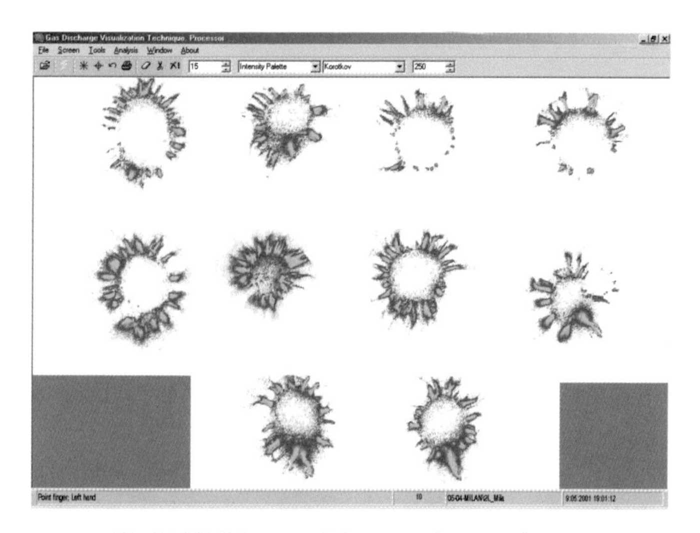

Fig.64. EPC images of sherpa Milan. Nepal 2001.

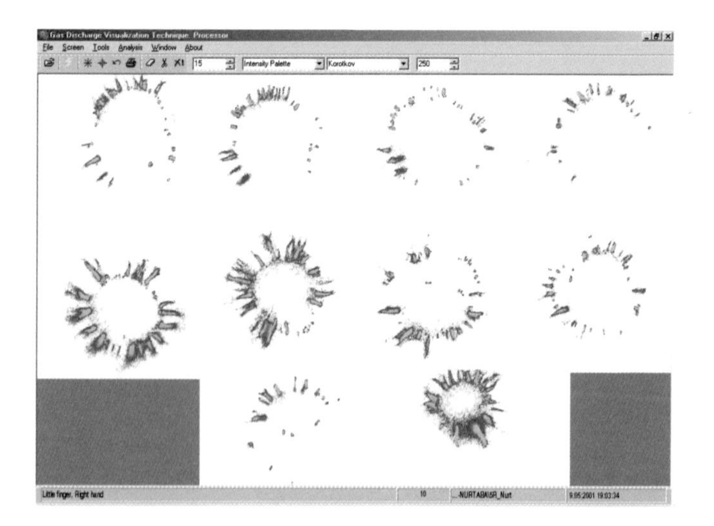

Fig.65. EPC images of sherpa Nurtaba. Nepal 2001.

JOURNEY INTO THE CONAN DOYLE'S LOST WORLD

The Adventures Begin

As dawns breaks on a frozen January morning, a mighty Lufthansa aircraft circled for the last time over a sleeping Saint-Petersburg, carrying our small expedition group away, first to Germany, and from Frankfurt airport across the Atlantic to Venezuela. A twelve-hour flight lay ahead. Having sunk into a comfortable leather chair, there was a chance to think about the trip ahead, this mysterious and secretive country, spread out on the shores of the Caribbean, highlighted mainly in green on the map, with the meandering blue strip of the full-flowing Orinoco River.

We flew into the Venezuelan capital Caracas as a whole group of scientists from St Petersburg and Moscow. The four days preceding the conference Systematic Medicine, Oncology and Bioelectrography were spent on the shore of the Atlantic, swimming on warm tropical waters, taking regular measurements, and discussing scientific problems with colleagues from the USA. This time was necessary in order for the body to adjust from the cold, polar Russian winter to the opposite: the eternally warm terrestrial equator.

Venezuela is close to the equator, although it is still in the northern hemisphere. So in January it was winter there. During the day, the temperature never rose above +30?C (850F), and the ocean water was 'cold' at +25°C (750F). The local inhabitants are not accustomed to winter bathing in 'cold' water, so we were the only visitors at the Atlantic beach club. We were glad that an American from northern Minnesota could join our party. The sight of our swimming heats made the Venezuelans shiver inside. Our attempt to bathe at night was cut short by a party of carabineers, who explained that night time was precisely when ocean monsters were partial to tasty Europeans. As in any Latin American country, it's hard to get by in Venezuela without knowledge of Spanish. Only the highly educated intellectuals speak English, although with some luck it is possible to come across Russian emigres in Caracas.

But our Petersburg-Moscow group was not just there to have a nice relaxing time in the Venezuelan Riviera. There was hard work to do at the congress, and then an expedition to the summit of one of the most mysterious table mountains of Venezuelan Guyana, Mount Roraima, situated in the heart of the hot Venezuelan Great Savannah, surrounded by humid tropical jungle. The ascent to the summit turned out to be quite different to the journey down the slopes of Everest, or among the Bezengi mountains in the Caucasus, or the excursion to the Solovetsky islands which had been led by Professor K. G. Korotkov. Despite the difference in expedition conditions, all of the expeditions had concurrent aims - research into human energy, research in the field of bioelectrography.

We would like to tell something of the expeditions themselves and their basic outcomes. At the same time, we must warn the reader that we didn't have any mystical encounters; we didn't uncover the Shambala or talk with magi. These were journeys into interesting corners of our little planet, which are theoretically within everybody's reach, motivation, time, and a certain amount of spare cash permitting.

And finally, our jackets and ties had been left in the hotel, and we were flying out of Caracas to the town of Puerto-Ordaz. Since we were spending no more than an hour in this glorious town, or rather its airport, we crammed into an air taxi - a small five-seater plane called 'Sesna' which looked like a model dating from the middle of the last century. It zoomed along the runway and the plane shot upwards into the air, whilst the green spaces of the tropical selva spread out beneath us. The plane wasn't flying high, perhaps at an altitude of 500 to 2000 metres, and the weather was ideal. And beneath us, stretching to the horizon lay a green carpet with the ribbons of rivers, the blue eyes of lakes and the ridges of the Table Mountains, overgrown with forests. It was a formidable sensation which could only be conveyed in words by a true craftsman, such as Saint-Exupery or Richard Bach. Passengers in large aircraft don't get to feel anything of the sort.

Of course, they don't get to feel the little pockets of air to which the 'Sesna' was so sensitive. So whilst most of our team were admiring the landscape, some were sharing the breakfast they had just eaten with a plastic bag.

From our cruising altitude, a view of the Guyana plateau opened up before us. It is a vast region to the south of Orinoco, covered partly by savannahs, partly by rainforests. With the exception of its north-eastern outlying area, the region as a whole is sparsely populated and undeveloped. Small tribes of 'forest' American Indians live in stream with the rivers, which they use as a basic means of communication. They hunt, fish, have primitive agriculture and collect wood. Until the Second World War, the only significantly populated area in the whole of this vast territory was that same town of Santa Elena, situated on the right bank of Orinoco, which was where we were flying to. It was the gateway for anyone heading for the deep Guyana plateau, the land of gold, diamonds, and rubber. But the discovery, in the north of the plateau, of extremely rich deposits of high-quality iron and magnesium ore, uranium, bauxite and other useful minerals, combined with the powerful hydroelectric resources and a convenient transport network meant that the region was ideal for commercial exploitation.

As we flew over Orinoco, we noticed only one long bridge. The bridge was incredibly beautiful; it was a delicate construction, hanging from two strong bearings. On closer observation, it turned out not to be a bridge at all, but an oil pipe line. The 'bridge' looked like a symbol of man-made European civilization, which, although discreetly, was nevertheless threatening to impose itself in the green body of the

Venezuelan countryside. It inevitably reminded of the Spanish influence, which began to manifest itself in the country's culture from the fifteenth century onwards.

From the air, the high waters of Orinoco appear truly grandiose, akin to the expanses of the closed sea. Many islands huddle together, reminiscent of parts of the human body. As you look at them, you cannot escape from the idea that the lines of ancient legends are coming back to life before your eyes. Many of the cosmogonical legends of the world's peoples say that when the universe was created, the great God-creator or Creator-Mother of the world created everything in existence from his or her own body. The body became the earth's mass, the eyes became the sun and the moon, hair became vegetation, blood became the water of fresh reservoirs, and breath was the wind.

For two and a half hours we flew over 600 kilometres of wild selva and saw the first signs of civilization: Santa Elena. It is a small town on the Brazilian border. A cluster of little houses in varying degrees of disrepair, the main street with plenty of little stalls, and mobile communication within the city. The population is a mixture of Europeans, Indians, and their descendants, a gaggle of children of all different skin colours. They are used to tourists, and the encounter with our jeeps did not arouse any excitement among the local population.

A visit to the outskirts, an evening in the delightful fazenda (ranch) in a tropical garden with parrots, affectionate dalmations, beer and pineapples. Tomorrow morning we are to enter Roraima.

The Grand Savannah and the tropical selva

The red earth road at first winds its way, ribbon-like, over nearly 30 kilometres of gentle slopes, before clambering to the summit of the plateau. The final point of the road is situated at a wattle and daub structure on the edge of a small village. A pole with the sign "information point" stood out clearly on entering the village. Here the National Guard looks tourists up and down - they are the inspectors of the National Park. A guide takes responsibility for the onward transportation of people within the Park territory, and follows strict instructions: "Lose as few tourists as possible on the way, otherwise there'll be no end of trouble". This is where the walking begins, for the next five days. The group of tourists, of whom there are rarely more than ten, as a rule, is served by 15-20 porters from among the local population, the Pemon Indians. The porters' job is not only to carry the heavy items and rucksacks of the tourists, but also to set up camps and prepare food. Each porter carries a load of up to 15 kg (33 lb) on his shoulders, packed into spacious porter-rucksacks called 'acau' in Pemon, which are set onto a frame made of palm branches, with straps from palm leaves attached to the forehead and strapped around the waist. With this baggage, the porter can make a swift passage over 3-4 hours to the temporary camp, where he awaits the tourists, taking time to prepare the night quarters and supper. The porters have a vivid appearance, with a loose vest, Adidas caps or straw hats with flaps, Bermuda shorts, beneath which ankle-length jogging bottoms can be seen, as well as trainers and the acau on their backs. Some of them can even manage the initial gentle slopes of the journey on mountain bikes. There were also women Pemon among the porters. All of the porters and guides were distinguished by their gentle temperament and their smiling, friendly faces. Evidently this is typical of the inhabitants of Latin America. (Fig. 66)

You may remember that in the novels of Jules Verne, Mine Reed and Conan-Doyle there is a description of expeditions into the unknown world, with a convoy of porters stretching along the sun-scorched track. Well that was exactly how our expedition moved forth. Seven Europeans, the first Russians to land in these parts, and 24 porters, Indians of the Pemon tribe. The path crosses the Grand Savannah, where rough grass turns green in the middle of winter and the equatorial sun burns every piece of European skin (mainly our noses and ears). Here the path climbs up the next hill, and before us the panorama of dark blue mountains opens up in the distance, and strips of jungle, or selva, on the mountainside. The beautiful view means you can make a stop under the pretence of a photo opportunity, and the selva beckons with the promise of coolness and water.Water in the Savannah is life, giving birth to lush green, hairy ants and innumerable birds. As you cross the border of shade, you leave the scorched world of the Burning Sun and enter the world of Green Madness. In the selva you can only move along the paths, either that or hack your way through creepers, brandishing a machete. The air is steeped in wonderful aromas, the local cuckoo taps away, as if hammering on rails, in the foliage someone rustles and rushes off. In fact there's only one beast to be afraid of there - the dreaded mosquito. At the start we hardly noticed the little black spots which clung to the skin near ponds. And it was only once we had felt the painful bites that the penny dropped - so that's who it was! Our epidemiologists back home had scared us with this, spoiling our health with vaccines. In comparison with these bloodsuckers, our native gnats seem like pleasant, doddery intellectuals, who warn you of their coming with a special signal (odious as it may be, particularly in the middle of the night). South American mosquitoes wasted no time buzzing around. They alighted without a sound and got stuck in. Painlessly and unnoticed. And then the skin was covered in spots, it swelled and itched. And this lasted for a week or so. It's a good job that none of team had an allergy. And no one got malaria.

But the real treat was in store for us at the end of each day of walking. You rip off your glasses and remaining clothes and throw yourself into the warm waters of the rapid rivers. Rio Tek, Rio Kukunan - these names sound like lines from an adventure novel, and they really do bring back the memory of the most hedonistic moments of our trek. 15 minutes in the boiling rapids, and you are ready for the next leg (or, even better, for dinner). Whilst you are splashing around in the water, you feel you are carrying out some scientific duty. For this is what we are researching: the influence of water on energy. (Fig. 67)

A strange feeling of isolation envelops the tourist who wanders along the mountain-savannah path. But it is only an apparent isolation. On the way the porters constantly outstrip you, giving a friendly wave, and the guides, always ready to come to

Fig.66. Pemon porter on the way to Roraima

Fig.67. Crossing Kukunan river.

Fig.68. At the top of Roraima.

Fig.69. At the top of Roraima.

your aid, watch over you from the neighbouring hill, tourists coming from the opposite direction greet you with a smile and the Spanish 'Hola' or American 'Hi', and the Russian expedition members wait at the next improvised rest-stop by a sharp bend in the path. The walking pace up and down along the path to the peak of Roraima is quite hard-going, so be ready to move those legs. For one of the members of our expedition, Professor Elena Okladnikova, those first few kilometres of the approach were exceedingly tough. To be sure, it was testimony to the long standing lack of training in mountain running (it was nearly ten years since the last expedition season to the high Altai Mountains. Our ascent to Roraima began on 25 January in the middle of the day. The path rose and fell, there were footbridges over streams which flowed in deep shady canyons, covered over by the selva.

Having surfaced from one such canyon, we realised that we would have to clamber along the stone track to the lower border of the Guyana Plateau, to the foot of an 800 metre (2600 ft) wall, marked in all geographical guides. We clambered up it in different ways: Konstantin Korotkov and Aleksei Khovanov zoomed onto the ridge, like two mountain eagles. Two other participants of our expedition group - Vladimir Kolyagin and Vyacheslav Zvonikov, researchers from Moscow - got up there without any particular strain, since they were on fine physical form and had had many years of special training, including yoga. Professor Okladnikova, meanwhile, having reached the ridge path, i.e. having exceeded the climb, looked rather like a fish which has been thrown onto the riverbank and is writhing in fatal agony.

"My poor little heart was thrashing like that of a dying bird in a dark, stuffy cage. It was simply impossible to breathe in the blazing sun and a 28°C heat. For a moment I thought: "I'm going to croak, here and now. What use is this to me? And anyhow, what am I trying to prove with this idiotic rock-climbing experience?" Different coloured circles swan in front of my eyes, and I could barely make out the faces of my Moscow colleagues who were waiting for me on the edge of the lower border of the plateau of the Venezuelan Guyana. Having taken pity on my plight, they cheered me up with the news that as far as the next camp the path followed the upper savannah, without very steep inclines and declines, and they also imparted the secret of special 'mountain' breathing. I realised the importance and usefulness of the secret that same day, when we had to overcome the next inclines. Indeed I can say in truth that had it not been for that secret, there would have been no sign of me over the next few days at the longed-for summit of the Roraima."

Having reached the temporary camp before dawn, we were pleasantly surprised by the fact that the porters and guides had already seen to everything in the sleeping quarters. After a filling meal with local strong drinks, we went out onto the little square where our tents had been put up. Above us was a dark indigo sky where a myriad of rough, unfamiliar stars were burning. The moon had not yet risen, and in the darkness of the ensuing night the vast and expansive Milky Way loomed up directly above us, shining in all the lustre and splendour of Orion, crowned with a belt and armed with a sword, surrounded by his hunting dogs, Venus was shining in the east, and over the edge of the table summit of the Roraima Cassiopeia was glowing. It was the entirely different splendour of the equatorial sky, unknown to northern inhabitants, and contemplating it took your breath away.

The second day was just as interesting, as we crossed a mountain river, scrambling along the slopes burnt harshly by the sun, and an unexpected swamp under the slopes of the Roraima itself. We continued the ascent over earth carved up by gorges, deep and rough streams. One of the strongest of these was the Kukenan, a tributary of the Orinoco. As it flows from the summits of the Table Mountains, the waters of the Orinoco tributaries is heated by the scorching rays of the sun, to the extent that as we swam across the Kukenan, we had the feeling that we were swimming in fresh milk. After the crossing there was another steep climb to the sheer slope of Roraima. This incline took up most of the second half of the day and we did it under a burning sun. We had to watch our feet carefully, so as not to twist an ankle on the sharp stone slabs. Add to that the dry, prickly grass of the savannah, spreading out into a miniature forest scattered about the savannah, not giving any shade, and the clay red soil, eroded after rain, of the path which ascended vertically up the mountainside and we have a full picture of the space which captured the imagination.

Under the scorching sun, we came to suppose that there was water above and below - which is the most natural state of the local terrain.

The Lost World

On the morning of the 27 January we were awakened by the sound of the rain. Or more precisely, it had been raining all night, its muffled drumming on the sides of the tent, but in fact it gave a sensation of comfort and coziness: it was pleasant to be drowsily aware that the flimsy sides of the tent were soaked and horrible, whilst inside all was warm and safe. But when the sound continues, and you realize that you have to go out in about an hour, that feeling of comfort instantly disappears. What does the wise guide say about this? It's time to get out your raincoat and drag yourself out into the street.

There were worse things to be feared: we are sitting in thick cloud, and you can't even see the mighty slopes around us, and a colleague has gone off to do business in the nearest bush. It lasts a very long time. The wise guide was none too pleased either: 'The tropical rainforest', that is, there will always be rain. There's nothing to be done: have a coffee, pull on your raincoat, and onwards and upwards.

It's true that not everyone from our group managed to overcome the pains in their joints and souls, which is why 6 people and only 14 guides carried on upwards.

Just after the camp the path climbs steeply up the Roraima wall. The ascent goes in steep stone steps, so our worst fears didn't come true: the way was horrid at first, but not slippery. And that was a good thing. But we would have liked to come down the sodden steps from top to bottom with cinematic cries.

The rain didn't stop all day and all the following night, but we had ceased to notice it after 15 minutes. All the more so since the raincoats which we had got in Petersburg at a high price were soaked through within the first half hour and it was only common polyethylene which saved us, as we had decided to wrap our rucksacks in it in order to preserve a few dry things. But, like any phenomenon, the never-ending rain had a flipside too. The little brooks and streams which flowed from the slopes of Roraima were transformed into powerful raging torrents, and seething waterfalls crashed down from the sheer five-hundred meter walls.

Venezuela is the land of waterfalls. The highest waterfall in the world, Angel Falls, is there, falling from a height of 1100 meters (3600 ft). Almost every river is abundant in verges and ledges, where the water falls from a height of 50-100 meters (150-300 ft). From our path we could see to the Tipui-Kukunan wall, where water was falling in a powerful waterfall from a height of more than 500 meters (1600 ft). It was an impressive spectacle, although an even stronger sensation awaited us, when before coming out onto the summit, we had to pass through on cliffs under the flow of a 100-meters (300 ft) waterfall.

You are of course familiar with the botanical gardens, with the tropical section. So just imagine the same garden made longer by a few kilometers, with sumptuous orchids, tall bamboo, and other tropical plants, with creepers twining around every available shoot. The ascending path to Roraima was full of such delights. It seemed surprising that such an ascent was even possible. And it really raised itself up into a sheer 600 meters (1800 ft) bastion. Somehow I found a book describing an alpinist expedition to conquer Roraima. The sportsmen climbed for several days, sleeping in hammocks slung from their alpinist hooks and soaked through with streaming water. One must be filled with fervor in order to be resolute about doing such a thing. Or, because of an argument. For us it was simpler. We went up steep inclines, along the path, along huge crevices cutting through the side of Roraima from top to bottom.

Threading our way through the bushes of the humid rainforest which crawls along the sheer slopes of Roraima, going right up to the sheer slope of its 700 meters (2000 ft) plateau summit, would have been impossible had it not been for the narrow, stony, in places flooded little path. Often stretching over hundreds of kilometers of forest, whole of the south-east Guyana is tableland. And it is there that the plateaus are covered with the 'lungs of the earth' - the equatorial Amazon forests. At the present time these forests are still untouched. Narrow paths for the most part run along the rivers. They bear the grand name 'routes'. As for motor routes, a single asphalt road cuts through the entire country - a motorway of international significance - which links Venezuela to Brazil. It is rare to find cars on the road, there are no markings, and in places the motorway is partitioned off by fallen trees which no one has managed to clear away. Drivers slow down, accurately driving round the natural obstacle in the opposite lane. Although it's a good thing that no Kalashnikovs appear from behind these natural obstacles. Had we been somewhere in neighboring Columbia or Bolivia, that would have been entirely to be expected.

And then we were at the last few meters of the steep incline, and coming out onto the stone platform of the upper plateau. We're at the top! Before our eyes there unfurls a surprising view: for hundreds of meters around there are black stone ledges, sometimes in the most fantastical forms, like bastions or abstract sculptures, magical structures. By association all that would come were the creations of that wizard from Barcelona, Gaudi, or Martian photographs transmitted by the latest interplanetary station. But this stone desert was full of life: each trough and hollow was filled with water, out of which jutted fantastical bushes with multicolored flowers. A real swamp of the Lost World; (Fig. 68, 69)

In this landscape, the appearance of a small plesiosaurus or pterodactyl would have been quite natural, they were all hiding somewhere. In terms of animals, our guides only found a little black frog which was croaking at me funnily in my palm. Yes, the fauna had reduced in the years since Conan-Doyle...

Having spent another hour with the stony swamp, we went up to an immense cliff, whose entire bottom half seemed to be a grotto, its hospitable roof overhanging the dry rocks. This was our campsite. A place where you could put up a dry tent, get changed into some dry clothes and have a shot of vodka, as offered by the attentive guide. How well we understood our distant ancestors, settled in caves, as they observing philosophically the raging poetry outside.

The next morning we were greeted by the sun. It became apparent that it was shining just over our plateau. All of the remaining space below us was covered in dense white cloud. We were, after all, at a height of 2800 meters (9000 ft) above sea level. And above the cloud level. As we went to the edge of the platform, we saw a white shroud of clouds puffing up at our feet. And you could only imagine how the cloud stretched downwards meters below. How interesting to jump off and fly through the clouds (although admittedly better done in a hang-glider).

We spent half of the day having walks around the plateau, examining the wonders of nature, and taking water measurements and measuring the fields of all of our expedition participants, including all of our porters. The results were really interesting, and even exceeded our expectations. The energies of fields and waters, measured with the help of the '5th element' sensor turned out to be significantly higher on the Roraima plateau in comparison with all other measurement points. And the time curves were particularly unusual!!! And that's without mentioning the fact that our expedition members demonstrated high energy values in precisely this area.

In this way, the idea of a special nature of energy of the zone of the Lost World and its influences on humans was fully confirmed. In our view, this influence should not be attributed solely to the emotional excitement or impact of the altitude. We took similar measurements in mountains in the Caucasus and Nepal, in the heights of the Crimea. There we also recorded the changes in energy among experiment participants, but it was predictable and correlated with their physiological condition: it decreased under a burden and increased after a deserved rest.

JOURNEY TO THE INDIANS OF SIERRA NEVADA

14 August 2005, 6.20 am. 1860 meters (6100 ft) above sea level, a mountain stream babbles nearby, gushing over stones. I'm sitting on a little rock beneath an enormous pine, watching as clouds of fog rise above the surrounding mountains in the rays of the emerging sun. The only trouble is the drops from the branches which periodically fall directly onto the pages of my travel journal. Columbia, valley of the Sierra Nevada. There was a tropical downpour at noon yesterday, but when I went outside at 2 am, the sky was covered with billions of stars, among which I could barely make out something faintly resembling the Southern Cross. From time to time, bright little stars flew into the sky - it had been officially announced that the Earth intersected with a strip of cosmic dust, and tiny meteorites were burning up brightly in the atmosphere. Meteorites are in fact no rarity in Columbia. It is said that a big meteorite fell in the neighbouring valley, narrowly missing a local dog, and the meteorite, weighing in at 22 kg (48.5 lb), is kept in the national museum of Bogota.

Yesterday we travelled four hours to get here from the little town of Valledupar. At first the road went happily by along the highway surrounded by cocoa plantations and green fields, but after an hour we left the tarmac behind and went up into the mountains. At that point the road ran out. For the road our Landcruiser had slowly taken was barely a road at all. At some point this track had been passed over by a mighty grader, cutting a relatively smooth road out of the humps, but it was clear that a lot of time had passed since then, and nature had steadily returned the construction to its natural state. Bursting streams had eroded the clay soil, creating deep troughs in it, huge stones rolled down from the surrounding steep slopes, and our Landcruiser crept along the edge of deep precipices which took our breath away with a sensation of dizzying depths. But our Indian guide calmly turned the steering wheel whist still managing to chat on his mobile phone, which he kept pressed between his ear and his shoulder - he wouldn't have been able to control the car with one hand. During three hours of journey we met three jeeps coming in the opposite direction, one of them, as is usual, in the most inconvenient place possible, and we rolled back a fair way to give way. The higher we went, the more frequently we met picturesque horseman in white clothing, prancing around sedately on business.

After the first incline, there before us was a green valley, surrounded by the gentle hillocks of the foothills. It was reminiscent of the crater of an ancient volcano, so perfectly-formed was the ring of foothills, surrounding the valley. We saw a group of neat hats with reed roofs, which immediately made me think of Ukrainian mazanki (daub cottages). And most surprising of all: oranges and lemons were growing in orchards next to a green maize field, small coffee plantations were the equivalent of our kitchen gardens, and above the valley the air was full of the wonderful scent of fresh pine. Huge scale-winged insects flew past, black suckling pigs busily rummaged in the earth, a parrot jumped in the branches, and the gaggle of poultry resounded from every village.

Indians have been living in this valley of paradise for the last two to three thousand years. And so it was that historically these places were never touched by great American empires - neither by the Incas from the South or the Maya-Aztecs from the North, and the Indians lived in the vast territory of the Sierra Nevada, out of reach of merciless conquerors. The Spanish went up into the mountains in the 16th century, but there had never been gold there and they lost interest.

The most surprising thing of all is that these tribes have managed to maintain their traditions and their autonomy over the last 500 years, in the face of pressure from modern civilization.

We climbed out of the jeep and passed a few small well-kept houses, and we were led along a path behind the village. Crossing the ford, we went up the hill and saw a large group of Indians in white clothing and white hats sitting under a high tree. They were waiting for us. All around stood women in white who were constantly knitting something and small children were running around. As a welcome, the chieftain gave each of us 4 white cotton threads - a symbol of what man gets from nature. The Indians extract these from the huge leaves of the cactus plant and then make clothes, bags and sacks out of them. The resulting cloth is quite tough and resembles jute. A leisurely discussion ensued. I spoke in Russian, Alexei translated into Spanish, and the chieftain, having listened to each long tirade translated the whole thing in its entirety into the local language. He then replied in excellent Spanish. Sometimes the Indians struck up a discussion; they spoke in turn without interrupting each other, as if thinking carefully at the end of each speech. All the while they chewed on coca leaves which they took out of special little sacks. The bowed to greet each other and exchanged coca leaves from their sacks. (Fig. 70, 71)

The Chieftain said that their tribe retained the traditions of their ancestors and the way of life which had survived here for thousands of years. They refused electricity and European tools and equipment. They deliberately don't repair the road to limit the influx of tourists and official persons. It has to be said that once you've travelled on that road you don't want to repeat the experience.

During our entire meeting the women were knitting, gaggles of lovely children were running about. It was clear from the number of children that evenings here are long and dull, and the tribe has certainly made provision for its future survival.

The exchange of opinions about the world and nature continued until three o'clock. When the chieftain announced the end of the discussion, it began to rain. This was warmly welcomed, as it was the sign of the benevolence of the heavenly forces. After that we were invited to eat with them: chicken soup with yucca and vegetables, and roast chicken with rice. As we had noticed, there were many chickens running around in the yards, but it seemed chicken was festive food for the Indians. A tropical downpour severely restricted our activities after that, and we

Fig.70. Meeting with Indians.

Fig.71. The Mamos.

Fig.72. Holy dancing.

Fig.73. GDV measurements in 'mamo' village.

spent most of the evening engaged in leisurely discussions. When dusk came we went to bed. When dawn broke, we woke up. Village life without electricity does have its charms all the same.

After breakfasting on papayas and pineapples, we took the EPC GDV from some Indians after we had been invited to a ceremonial hut. Walking through the village, we crossed streams on stepping stones and a narrow path brought us out into a clearing where there stood two reed huts and Indians sat in groups. They were ceremonial huts - one for women, one for men. Inside, the construction was made of black poles intertwined in a complex manner, with central pole 5 metres high. From the outside the cone of the hut was crowned with a pillar construction resembling a modern antenna. The Indians said that this ancient construction obtained energy from the sky, which then went into the hut via the poles, where it was received as a constantly-burning bonfire. Another argument in favor of the theory about contacts with other planets. Smoke kept on puffing out of the wicker rood and entrances.

The dancing began. The women began to turn on the spot to the sound of the drum and pipe, now together, now in pairs. The pace of their movements looked like a tipsy 60-somethings party. (Fig. 72)

Then we went to the male hut and began to measure the GDV of 'Mamos'. Let me say a few words about who they are.

There are about 20 000 Indians living in the vast territory of Sierra Nevada de Santa Marta. They settle in small villages, sometimes as separate families. The abundant natural surroundings, a constant temperature of 20-25°C (65-75°F) throughout the year and the lack of predators and malarial insects create ideal conditions for life in these mountain areas. Moreover, such conditions are conducive to a leisurely, lazy life which doesn't demand particular effort or any struggle for existence. There are no wars, no newspapers, and no big cities. There is no need for written language or craftsmanship. Dried pumpkins serve as plates; clothes are made from cactus fibres. Exchange between territories has always been difficult, as it is difficult to get around the mountain paths on foot, and horses only arrived with the Europeans. It is a tribal system which has survived throughout the centuries.

The organization of life was taken care of by the institution of the 'Mamo'. In the local language this means 'the wisest ones'. They are chosen from among the brightest youths so that the older Mamo can pass on to them the traditions and customs. The Mamo grow up, get married, have 8-10 children, and live the lives of normal people. People turn to them for advice, to resolve disputes or to take decisions which are important for the tribe. A few times a year, all the Mamo meet in a central village, Nabusimake, where they sit down together and hold a leisurely discussion whilst chewing on coca leaves. In the special place, under a sacred tree, the Sacred Spirit descends to them. (Fig. 73)

The Columbians we spoke to in Cartagena and Baranquilla believe that the Mamo have special qualities. They can foretell events and fashion the world to their will. As an example, in spring the Mamo had already announced that scientists would come to them in the summer, and that they would receive them. The Mamo are very selective in their contacts with civilization and it is practically impossible just to turn up there.

The Indian Sierra Nevada is striving to maintain its autonomy, and the Mamo are the spiritual leaders of that process. They protect their people from the influences of Western civilization, and in many ways they are succeeding. Most of them, including young people, wear national dress and uphold their traditions. Of course, Catholic missionaries have been to the Mamo territory, and they built churches there and began to preach. The struggle continued for many years, but in the end, 23 years ago, the Mamo managed to drive the missionaries out. The church doors were closed, and the Indians worship the spirits of the mountains and rivers as they did a thousand years ago. At the same time, they are quite happy to use mobile phones and many of them wear boots as well as reed sandals, as they are more comfortable in the rainy season.

Of course, civilization edges ever closer: metal pans are more convenient than clay ones, children nibble away on biscuits and sweets, and the local arrack is poured from plastic bottles, to say nothing of the horse saddles and harnesses. Some of the Indians go around in jeans and wide-brimmed hats.

The major problem is health. The government opened a medical centre, and a friendly doctor provides first aid, just as in a normal, small rural hospital. But naturally, in more serious cases, they are at a loss.

We examined a few patients. The illnesses were typical of rural life with a lack of hygiene. Intestinal parasites, the consequences of the past injuries, childhood diseases. The lack of decent road access means that one can only trust in the forces of nature. It is interesting that no Indian knows his age. In a world were there are no changing seasons to speak of, life flows by as a single calm river.

The Mamo asked us to convey their message to society: 'Allow us to live our own life. We have managed to preserve our traditions for thousands of years, allow us to continue to live in this way. In our settlements, Indians live in the way that they are accustomed to; they are happy and raise their children in peace. There is no theft or crime here; people live honestly according to the laws of their ancestors. We do not need cars or televisions, which go hand in hand with unruliness and drunkenness. We chew on cocoa leaves and that is enough for us. We believe in our spirits, and they have protected our life for thousands of years. Leave us in peace and let us live our lives'.

After long negotiations, the Mamo managed to agree with the Columbian government about its autonomy. A black line was drawn which confined the territories of the Sierra Nevada Indians. But the Mamo complained that the government constantly violates these borders, and nobody in the government wants to listen to what the Indians are saying. They have enough on their hands with drugs cartels and guerrillas.

The Sierra Nevada Indians are present on the international stage too. Their leader Rojelio Mekhia was invited by UNESCO to participate in the 2004 congress on tribl culture in Tokyo. He didn't like the city: 'There are too many people, it's noisy, veryone's running off somewhere'. In the autumn he intends to go to a congress in ;pain. It should be easier there, since at least it's in Spanish.

We left the Indians' village in the middle of the day. Storm clouds had already larkened the sky, and the return journey was not expected to be a happy one. In fact, ve had to dig the jeep out of deep clay holes, and the skill of the driver sliding around n the sodden road above the precipices made the heart skip a beat.

The Mamo smiled as we took our leave, and clapped us on the shoulder. We went ack down to civilization to our worries and problems, and they remained in the vorld of pure energies, babbling streams, and good-natured smiles. May they preerve their way of life the way it is! Humanity is developing new technologies, xploring space, killing in wars, drinking hard and frittering away valuable time in ront of the box, and somewhere in the mountains of Columbia the Sierra Nevada ndians summon the rain by tapping on a turtle shell. It is in this diversity lies the reatness of the human spirit, the possibility of constant development and renewal. This is why it is so important to retain the autonomy of cultures, to preserve their ndividuality and in every possible way to support the effort to avoid Western stanlardization. The more diverse the world, the richer the human spirit! (Fig. 74).

Fig.74. The Beauty of the World.

MEASURING GEO-ACTIVE ZONES WITH "5TH ELEMENT" SENSOR

The problem of registering geo-active zones with objective instrumental methods is one of great scientific and practical significance. The widely employed methods of biolocation make it possible to obtain interesting information whose value can be greatly increased owing to instrumental verification.

Formerly, we developed a gas-discharge sensor which gave interesting results during the registration of the distant impact of the human operator . My current work describes the construction and results of trials of a GDV sensor working in a set with GDV instruments.

The principle of the Five Elements is one of the bases of Traditional Chinese Medicine together with Yin - Yang principle. Theories of the Five Elements emerged from an observation of the various groups of dynamic processes, functions, and characteristics observed in the natural world. Each Element is seen as having a series of correspondences relating to the natural world and to the human body. Fire, for example, corresponds to Heat and to the Heart. A pattern of interrelationships between the Five Elements is used as a model for the way in which the processes of the body support each other. These are defined mainly through the Sheng and Ke cycles (Fig. 75). The principle of Five Elements was used in the construction of a new sensor.

Fig. 76 shows the principle of the sensor.

The GDV Camera Pro or GDV Compact Camera serves as a measuring device. Titanium calibration cylinder is placed on the optical glass of the Camera using special holder. Camera should be run on 12 Volts batteries - this is a very important condition!!! Special automatic connector allows connecting the wire of the cylinder to 5 contacts. In the first case cylinder is connected to the Camera (contact 3 of Fig. 76). Other contacts allow connecting to different sensors: electrode placed in water (river, stream, lake, etc.); electrode grounded to the earth; elec-

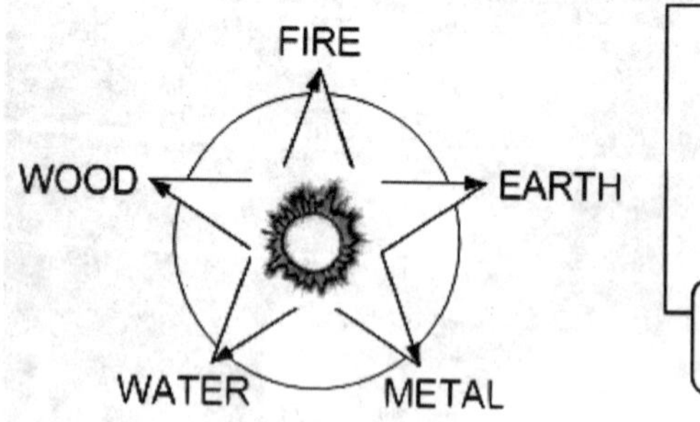

Fig.75. The Five elements principle.

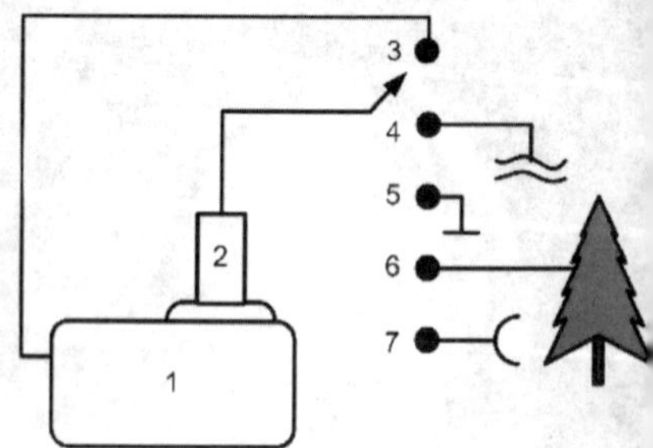

Fig.76. The principle of the "5th Element" Sensor. 1 - GDV Camera; 2 - titanium cylinder; 3-7 - sensors.

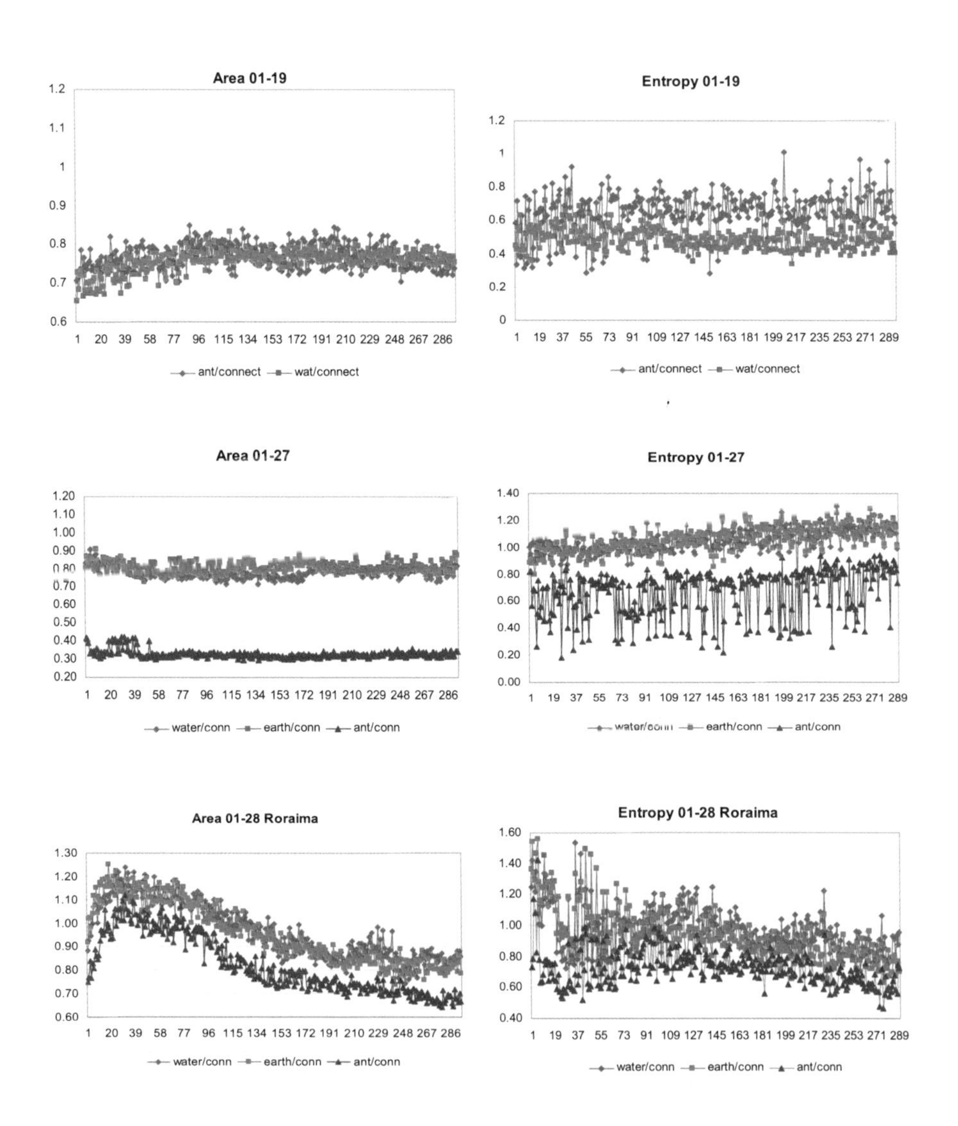

Fig.77. Examples of time dependence of Area and Entropy.
At the see level on 01-19 both Area and Entropy demonstrated quite stable signal.
Nearby Roaraima 01-27 the signal of Entropy from Antenna sensor became
very un-stable, while at the top of Roraima 01-28 all the sensors demonstrated
very high variability both for the Area and for the Entropy.

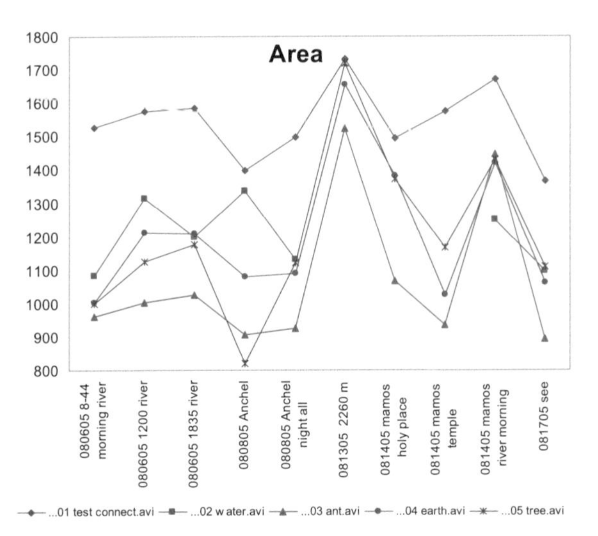

Fig.78. Sensor signals at different days. (01.avi) - connected to the GDV instrument; (02 water.avi) - water sensor; (03 ant.avi) - antenna; (04 earth.avi) - earth; (05 tree.avi) - tree. Venezuela - Columbia, August 2005.

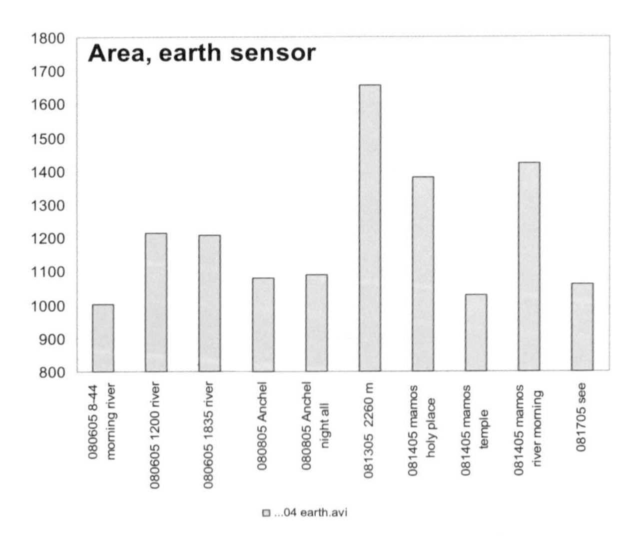

Fig.79. Earth Sensor signals at different days.

trode stick into wood and electrode connected to the antenna. At every contact position dynamic AVI signal for 10 seconds is being measured.

A series of measurements were done in the process of expedition to Venezuela to the area of Roraima - Kukunan Tipuy (Mountains). Measurements were taken with the titanium cylinder connected to the following sensors:

1 - connected to the GDV Camera;
2 - electrode in water;
3 - antenna directed to the North;
4 - earth grounding;
5 - electrode stick into wood;
6 - electrode touching human body.

Examples of time dependences of experimental parameters are presented in Fig. 77. In normal conditions the line for Area is relatively flat and variation is less than 5% (fig. 77, 01-19). Strong trends were recorded on Jan 20 measuring see water, on Jan 28 at the top of Roraima (fig. 77, 01-28) and on Jan 29 near the Kukunan river for all sensors. (Fig. 77)

Entropy lines demonstrate a more complicated behavior. In several cases the dispersion of Entropy for the Antenna sensor was very high (Fig. 77-01-27, 01-28). It was on the days of full moon and strong changes of weather - on Jan 26 it was sunny, but on Jan 27-28 it was raining for the most of the day. Jan 29 heavy clouds were in the sky. We should remember that on these days measurements were taken at high altitudes - from 1960 meters to 2700 meters above see level. Measurements taken at 1000 meters altitude on a sunny day on Jan 30 demonstrated a "quiet" time line.

Measurements taken in August 2005 during expeditions to Venezuela and Columbia, demonstrated very interesting data as well. As we may see from fig. 64, the highest peaks were registered 08/13/05 and 08/14/05 in the Sierra Nevada area. This is more clear presented in graphs of Earth Sensor in Fig. 79. (Fig. 78, 79)

After processing all the expedition data the main conclusion was made that the "5th Element" sensor reacts to the natural energies of different places. A lot of research should be done, but we are sure that it may be very useful in selecting beneficial position for buildings, study of sacral places and research in geo-active zones. We hope a lot of researches will participate in this process.

EPILOGUE: WHERE DO WE GO?

"Every day of man's life represents a little piece of evolution of our species, which is still in full swing"
E. Shroedinger "Mind and Matter""

When we speak about future forecasts, all Humankind is divided into two big groups: Pessimists and Optimists. No matter whether the topic is about fishing on Sunday noon or perspectives of global survival. This division is independent of current information. This is a quality of brain organization, most likely, domination of one or another hemisphere. It is well known that the left brain is an optimist, and the right one a pessimist.

By nature I surely belong to the Optimists and believe in the progressive spiral of development. These views are supported by a whole life experience, the practice of mountaineering, a belief in life after death, and analysis of the history of development of human civilization known to us.

During thousands of years pessimistic Prophets regularly predicted the End of the World. This took place on the eve of every new millennium, and sometimes even a new century. The present epoch was not an exception. And so, the prophets disappeared, but human beings are still living. In the beginning of the previous century the pessimists' serious prediction that in forty years, because of rapid development of transport, London would be covered with horse-dung - this did not come true. Malthus proved by calculation that in the middle of the Twentieth century Humanity was threatened with total starvation - again a bad shot. We can gather a whole volume of such examples, so there are grounds for optimism. Let us live some more time. Question: how? Will we turn into a society of computer game players, controlled by the electronic brain, or will we destroy all ecology by genetic experiments and ozone holes?

There are many scenarios of development. And they should undoubtedly be calculated. First of all, not to waste money on the projects of Star Wars and biological cloning. Modern synergetics allows to build probability models of development and making certain forecasts on this basis [25]. Development of powerful computers will significantly contribute to such investigations.

At the same time, let us look at the processes of Humanity's development from the global, universal historical viewpoint. We can review the period of about 5000 years known to us, although there are some grounds to suppose that highly developed civilizations also existed before this period. The American researcher Graham Hancock presents a lot of interesting ideas on this topic in his books. [26]

So, if we consider the course of historical development as the process covering different nations and continents, developing in waves within centuries and millennia, we will witness a clear forward spiral tendency. Several aspects of this process can be distinguished: material and technical, mental and ethical, social, ecological, spiritual. Progress in the material sphere is obvious and needs no comments: we moved from the caves to the skyscrapers and slowly move into cosmic space. In the mental and ethical sphere the progress clearly manifests in the attitude to value each individual life, regardless of the social status of its owner. And this progress goes very quickly in various places of the globe, although in different ways. Some 200 years ago slavery was a common thing in Russia, USA, Africa, and Asia. Today we are fighting against racial, sexual, and age discrimination, actively put into practice a moratorium on the death penalty, and develop methods for rehabilitation of mental.deficiencies There is direct evidence of progress. In the sphere of ecology Mankind has changed from the conquest of nature and mass destruction of all life to the protection of nature and concern for dumb animals. In the social sphere more and more countries gradually transform from one social formation to the other, more complex and providing the highest level of mental and ethical development.

And, finally, a distinct tendency to the development of global perception of the idea of the Universal Spirit can be traced in the spiritual field. The lowest level - worshipping the powers of nature: thunderstorm, rain, wind, and sun. The next level - personification of spirits associated with these phenomena, giving them human features while keeping their super-human status. Then - creation of an assembly of gods and imparting all human weaknesses and habits to them, with a detailed description of their life stories, situations, and adventures. In other words, projection of the everyday life to the divine life. And, finally, the emergence of monotheistic religions, in various parts of the globe, in different cultural and historic situations, but in a relatively small historical time period. And these religions spread around the Earth, millions of people speaking different languages and belonging to different nations became involved in this process, and it was the next stage of spiritual development of Humanity.

Naturally, these processes occur unequally, they often return to the previous stage or come to a stop lasting for many centuries, however in historical perspective the development is obvious. An impression is created that Humanity develops as a single organism, gradually growing up and transforming from one phase to another. Here we follow the line of ideas developed by E. Le Roy [27] and V. I. Vernadski. Naturally, when we say "organism" we mean, first of all, collective consciousness, as a function of complex structure, consisting of a quantity of elements. This organism went through the period of naive childhood, pugnacious youth, and only now is nearing the maturity. Still very far from wisdom. Development of the organism of Mankind is an inevitable process, although often flowing quite unevenly. However, all the deviations from the total progressive line are temporary and after a number of over-oscillations the system comes back to the given mechanical trajectory again. This thought can be formulated as a theorem.

Theorem of Development

Progress of Mankind moves along an ascending spiral, as a self-organizing process of development of a single organism. All deviations from this process are temporary over-oscillations, which gradually come back to the main line of development.

To all appearances, these processes involve global powers, strange and incomprehensible to us, bringing historically significant events. History gives many examples of this. England, progressive from a historical view, was dreadfully threatened with the Spanish invasion, and it seemed that no forces could stop the "Great Armada." However, suddenly a very strong storm started, and all the ships of the Armada were lost in a waste of waters. Another example. While Napoleon was introducing the new progressive ideas in Europe he was invincible. When he started the aggressive war, blinded with his might, he was thrown down and defeated. All the Twentieth century was a period of massive historical experiments in the sphere of Human Spirit. The Empires of Stalin and Hitler represent global experiments in a forced creation of a New Man. Experiments which took away lives of millions of common people and ended with a total collapse.

Now we can observe the next phase of the global historical process: creation of a single economical and political European institution. This is a great step forward on the way to the creation of a World Government. However, this is an issue of the quite distant future.

The level of development in the modern world is very non-uniform. And very often we do not even understand this. Most people in Europe, Russia, or USA assume that there is a civilized world, where we live, and there are separate wild tribes somewhere in the wilderness of the Amazon. As a matter of fact, this is far from true. "West is West, East is East, and they can never converge", wrote Rudyard Kipling more than one hundred years ago, and this idea remains true even today, despite tremendous progress in a lot of countries. I happened to travel a lot around the world, firstly among mountaineering expeditions, now more and more often with lectures and workshops. I visited far places of the Middle Asia, particularly the borders of Afghanistan, India, Nepal, and, what is more, far away from the tourist tracks. And each time I wondered how different is the life-style of the local people from our Western standards.

Two days by truck from a large city, and you find yourself in the medieval world, living the same slow life and following the same rules as thousands of years before. People graze cattle, wandering with the herds in the desert mountain gorges. They marry, give birth to children, grow old and die according to the ancient tribal traditions. And it is not important that they have a refrigerator or a TV in their house, that in one or two days they can find themselves in a big city with all the attributes of civilization - this is another world for them, alien and needless. Not separate villages, but whole countries and nations, most of the population of the globe lives this life.

Therefore, when we come to them with our traditions, our culture, this evokes counteraction and antagonism. A paradoxical situation occurs: Western civilization distributes its equipment around the world ? from machine guns and rockets to computers and CD-players. This equipment is demanded and consumed in the whole world, regardless of the nationality and culture. But together with the equipment certain cliches of Western, first of all the American style of life, are spread out. But do they reflect the highest achievements of Western civilization? Can McDonalds or Hollywood thrillers pertain to the cultural progress of Mankind? This is a certain substitute, cheap chewing gum, but these are the products, which are spread most quickly. And these very products give birth to the highest antagonism, contradicting local customs, rules, habits, and evoking irritation with the "satisfied" American reality.

And this is, to my mind, one of the reasons for the present geopolitical situation. The West invests money, technology, weapons in the third world; all this is consumed by the ruling clique and, under conditions of feudal society, in no way influences the life of a majority of the population. Moreover, it is profitable for the ruling clique to keep this majority in ignorance and darkness. The deeper the feudal Middle Ages, the less information from the outer world ? the easier to keep people under control. Taliban supremacy in Afghanistan s the clearest example of it. Forbidding TV, radio, Internet, stripping the women of all rights - and the country falls under total control. And then turning the anger of the people against Americans, unleashing a "holy war".

What is the way out? And does it really exist or will we wallow in the meaningless war between the East and the West? The answer to this question follows from the concept of development of the Organism of Humanity as a single whole. In the present historical moment, strong non-uniformity of development of countries and even continents is evidently observed. In addition, the difference consists not in the quantity of material goods per head, but in the quantity of information. Humanity develops at the expense of information consumption. Hence, if we wish to reach a balance in the world, stable balance of negative and positive powers, we should distribute information in all ways. What kind of information? Positive and humanistic. What does this mean?

This information is formed of two components: individual and collective. On the individual level - access to education, mastering new knowledge and new technologies. If young people have the opportunity to study, develop, comprehend new horizons of life, they will contribute to the development of the whole society, its broadest masses. But will a herdsman be able to make a computer engineer? Naturally, each one will not be able to, and will not wish to. But there will always be girls and boys applying all their skills and strengths to meet the Twenty-first century requirements. And they will pull the whole society along. A vivid example is India, supplying the whole world with the best computer programmers and rapidly developing itself.

Collective information suggests the propagation of humanistic ideals, notions of the value of each individual life, spiritual equality of all people. These ideas are common to all races, nationalities, and religions, and their propagation is the only guarantee of the development of the Organism of Humanity.

So, if we wish to live in peace and rest, we should spread not weapons or coca-cola, but computers and books. Free information exchange coupled with a careful attitude to national features and traditions is the only way of development for Mankind. Therefore, the Internet and the satellite system of its support can be called the greatest achievements of the Twentieth century.

Information is the blood flowing in the veins of the Human Organism and providing its existence. Congestion in its parts causes diseases, and these diseases influence the whole Organism. The more actively the informational blood flows in the veins, the healthier and stronger every cell and every organ. And only now we understand that the Informational Society is not just a metaphor, but the only condition of further development.

Understanding the evolution of Mankind as an evolution of an individual specific organism enables us to foresee many tendencies in its development. Considering that we only now enter the period of maturity, having passed childhood and youth, it becomes clear that Humanity still has a long way to develop. All our mistakes and sins are an inevitable stage in the process of maturity. Speeding-up the evolution is a natural process of changing inner time of the system in the course of development.

In childhood time goes slower: from one birthday to the next the whole life passes by. With aging, the inner time increases - days, weeks and months fly faster and faster until the old age when a person cannot follow their sequence. This is not just a feeling, but the change of speed of inner processes, modification in the exchange of entropy with the outer world. The speed of exchange with the outside falls, the role of inner processes responsible for information processing increases, with the formation of information having new qualities, and inner time speeds up.

This process takes place in the human society. We lose direct dependence on the environment, start to produce more and more of our own information, and the speed of our inner processes increases, i.e. the inner time of the system decreases. This represents the beginning of a new phase of development of human society as a single organism.

Thus we can not be afraid of the pessimistic forecasts. We are just in the initial phase of development, in the first pages of the Book of the History of Humanity. As any organism, our human organism undergoes crises and diseases, but this is a temporary and transient phenomenon. Mankind is a self-organizing system, and after all deviations it comes back to its optimal way of development. In other words, the Organism of Humanity is a stable system, interacting with the biosphere and flows of cosmic information. So, we are not threatened with global catastrophes, although we will still be able to think out a lot of local problems for ourselves.

Modern mathematical concepts, based on the synergetic ideas of a theory of nonlinear dissipative systems enable us to build beautiful models of the notions given above. The concept of entropy, as a measure of structural system organization, plays the leading role in these models. The meaning of entropy widens far beyond the scope of thermodynamic approach in the theory of information and synergetics.[28] The EPC technique presents a mathematical approach to the calculation of entropy of individual systems on the basis of glow images - EPC-grams[29] .

This approach opens absolutely new perspectives in the investigation of behavior of the organism and its connection with the collective informational field.

As any other organism, the global human organism consists of separate cells. Each of these cells is represented by an individual person. The organism can exist only in the case when the larger part of its cells is healthy and functions normally. In addition, the more of these cells, the more complex is the system. And the more active the information transfer between these cells, the more labile and live the system.

Therefore, the first and the main aim of life of any person, any individual is life itself. For it is the individual life only that provides the rescue of the organism of Mankind as a whole. And the life of a street cleaner is not less important than the life of a member of the Government - they just have different specialization in the system of the organism. Thus, after this discussion, from the global perspective we have approached the thesis of the value and self-sufficiency of any human life. It is obvious that such discussion can be applied to the biological life in general.

"So, what about our individual development? Is there any progress found, and how do we differ from our historical ancestors?"

Individual development goes on, as well, but not according to the Darwinian variant and not in the physical body. Physically we in no way differ from our ancestors who lived some tens of thousands or, perhaps, millions of years ago. A Cro-Magnon man, dressed in a suit, would not differ from the crowd of passers-by on the streets of Manhattan. Therefore, we can offer two theorems:

Theorem of Physical Evolution

Evolution of the physical body of modern man reached a stationary level, i.e. stopped.

Theorem of Evolution of Consciousness

Evolution of modern man takes place on the level of individual and collective consciousness and is expressed in the formation of more and more complex informational structures, passed on from generation to generation at the expense of techno-cultural fields and education.

There are no doubts that modern children master sciences much quicker than their contemporaries 100-200 years ago. Six-year-old children already read well, however prefer their mothers to read, whereas in medieval Europe reading was a great art and even the majority of the aristocracy was illiterate. From earliest childhood our children receive a flow of information, their consciousness processes it, masters and understands, and somehow includes it all in their inner map of the universe. The model of an outer world created in the brain of a modern child is surrounded by a mass of complicated elements, which impart the complex character to it. A child organically perceives the TV, radio, books, and the model of the world goes far beyond the family cell.

When the European travelers got to an African tribe, living deep in the jungles, and were telling about their country for a long time, after the night spent with them the sachem of the tribe said to his tribesmen, "The country of these people is far away, behind the big baobab tree at the third bend of the river". To him the space of the Universe is limited by the hunting expanses of the known jungles. To our modern children the idea of other countries and continents organically fits into their picture of the outer world.

Powerful informational flow does not pass without leaving a trace. It forms physical structures of neural nets in the brain, which is the basis of consciousness. The more complex this structure, the higher the level of intellectual development. Formation of neural nets takes place in childhood and young age, grown-up children can use only structures already created.

Thus, we come across the phenomenon of intellectual, ethical and spiritual progress of human beings as a species, born as a result of transfer of all the information accumulated by Humanity from one generation to the next. We can call this process "Imaginary Lamarckism" after Erwin Schroedinger. The gained skills are not imparted genetically, and our babies are as helpless as thousands of years ago. But under the influence of the informational medium of modern civilization, the complex structure of the brain is formed, providing the functions of consciousness.

The present hypothesis does not need to be substantiated with a supposition of some special fields, transferring information. Such concepts are of interest to us, for example, the concept of morphogenetic fields of Rupert Sheldrake [30] and, what is more, we consider them to be a real factor of morphogenesis. This idea is further developed in our work. [31] However, these notions are not the most important for the concept of the evolution of a consciousness of one Humanity discussed by us. As seen from the above, it is enough to speak about the informational medium, created by the customary means of human civilization.

Naturally, formation of a complex structure of neuron nets requires much time and much effort. This is why our children should study from early childhood till they become adult. The Society shall be responsible for them. Of course, we must not make drones of them, but we should give them time for free development. They should feel what life is, see other countries, fall in love and make their own mistakes. All this is a part of the evolution process. The society should take care of them during this time and assign special means for their free development. School and high school play an important role. Russia has a centuries-old tradition of good education, our graduates take good positions in many countries of the world. Therefore, it is necessary to support this tradition on the state level and develop it, support the process of evolution of Humanity. We see more and more young people in modern Russian society, interested not only in primitive material well-being, but in the development of all society, development of Russia as a great state, accumulating the strict rationality of the West and broad philosophical ideas of the East. And our future consists in these new powers.

Does this enable us to speak, however, about a new stage in the evolution of Humanity as a species? Yes, of course. There are a number of definitions of the notion of "Evolution":

- A constant process, where something changes, passing to the another, usually a more complex or better form.
- In Biology:

  a) a theory of evolution, studying changes of groups of organisms in time, mainly as a result of natural selection, when the descendants differ morphologically and physiologically from the ancestors;

  b) historical development of connected groups of organisms.

Say we can trace the evolution of horses from their ancient ancestor, more similar to a dog, through a number of fossil individuals, significantly differing from one another morphologically, to the modern well-known horse. And if we now compare various breeds of horses, say, trotters and heavyweights, they belong to the same species regardless of all their differences.

If we compare a modern individual with his far-away ancestor tens of thousands of years ago, we will see that there is no morphological or genetic difference, in spite of the wide variety of races, nations and types of individual constitution and physiology. However, a modern individual differs greatly from the intellectual viewpoint. And this intellectual difference is determined by certain morphology of organization in the brain tissue structure. This structure is not fixed genetically, i.e. we do not hold to the Lamarck's theory of the inheritance of the acquired characteristics. This neural structure is formed in childhood, through the effects of cultural and informational influences.

The level of intellect of a modern child can be measured by quantitative, objective tests, and we see that it grows even within the known period of several decades during the Twentieth century. We can imagine comparison with the intellect of a child some centuries ago. Ideas that were previously taught at the Universities are now studied in schools.

Thus, we can realistically speak about an informational evolution of Humanity. Evolution of the Spirit, not physical body. This process continues within, at least, some few millennia and becomes quicker and quicker in recent years. What can we expect from this process?

If we extrapolate modern tendencies, we can assume that Humanity moves to an understanding and control of the mechanisms of interaction between consciousness and the material world. Consciousness will be more and more involved in the creation of our world in the course of evolution. This process goes together with the development of intellectual equipment, and imperfection of individual physical qualities will be compensated by the development of technical devices in many spheres. Consciousness and equipment will integrate. Artificial organs, controlled by the power of thought, overcoming sensory defects: blindness, deafness, inborn dysfunctions. Free mastering of several languages, ultra-speed digestion of information, capability to flexibly operate huge data flows. What is more, operation on the basis of intuitive associations, not hard algorithms, as contrasted with computers. A computer will simply become a part of the cultural sphere, an integral part of life, as jeans and T-shirts, providing performance of many service functions in a background mode.

And most important - a man will learn to influence the development of the outer world consciously, by the force of thought. And not only our world.

We can forecast a number of consequences of this informational evolution of Mankind:

Control of diseases and collective informational correction of the energo-informational status of an unhealthy person;

Overcoming of sensory defects by way of mental training and using technical devices, transferring information directly to the brain;

Development of new methods of super-effective digestion of information;

Mental journeys and conscious acquisition of information during sleep and meditation;

Mental control of natural processes;

Mental control of the course of development of social processes, individual life, social environment;

Conscious guidance of the course of historical events.

It is easy to see that all the mentioned characteristics are stipulated not by the appearance of particular, even super genius people, but by the evolution of society as a single whole. All these are characteristics of collective informational processes, synchronous acts of large groups of people, possessing highly developed individual consciousness. And again we speak about the unity of individual and collective evolution, about the development of Humanity as a single organism. The organism, providing the intellect and consciousness of the planet Earth and Universe as a whole.

REFERENCES

[1] Rudnik V. A. Geocosmic factor as a determinant of bio- and ethnosphere of the earth. Proceedings of International Congress «Science. Information. Spirit». 2000
[2] Kirlian D., Kirlian V. In the world of miraculous discharges. 1964, M., Znanie
[3] Korotkov K. Human Energy Field: Study with EPC Bioelectrography. Backbone Publishing. NY, 2002.
Measuring the Human Energy Field: State of the Science. Ed. R.A. Chez. National Institute of Health, Samueli Institute, Maryland, 2002.
Measuring Energy Fields: State of the Art. GDV Bioelectrography series. Vol. I. Korotkov K. (Ed.). Backbone Publishing Co. Fair Lawn, USA, 2004. 270 p. ISBN 097420191X.
[4] Bundzen P.V., Zagranzev V.V., Korotkov K. G., Leisner P., Unehstahl L. E. Complex bioelectrographic analysis of mechanisms of the alternative state of consciousness. J of Alternative and Complementary Medicine, v.8 N 2, pp. 153-165
[5] Anufrieva E., Anufriev V., Starchenko M., Timofeev N. Thought's Registration by means of Gas-Discharge Visualization. Proceedings of the International Conference. University of Ljubljana. Slovenia, 2000.
[6] Nicholas Roerich (1874-1947), the renowned Russian artist, left over 7,000 paintings which are in Museums all over the world. He wrote the libretto and made settings to the original production of "Le Sacre de Printemps" and instituted the "Master Institute for the Arts" in New York City. He also created and promulgated the "Roerich Pact and Banner of Peace" which was signed into law by Pres. Franklin D. Roosevelt on April 15, 1935 and which earned Roerich a nomination for the Nobel Peace Prize. It is a treaty to protect the creative works of humankind during times of war.

Mme. Helena Roerich's (1879-1955) writings have long championed the cause of Woman and the valuable achievements of the Feminine Principle in life. She sees the new Era as the Age of Woman, so needed to restore a lost balance to this planet. Her metaphysical works pick up where her compatriot, Mme. H. P. Blavatsky, left off and include two volumes of letters, which are commentaries on the Teaching of Agni Yoga, and various not translated manuscripts which are now in the process of development force showing itself in our material world. Hence, making an effect on this world.

[7] Ban'kovsky N. G., Korotkov K. G., Soloduhina V. A., Shigalev V. K. Some peculiarities of formation of gas discharge images under reduced pressure // J. of Technical Physics. 1980.V. 50, N 10. P. 2015-2017.
[8] N. G. Ban'kovsky, K. G. Korotkov, N. N. Petrov. Physical processes of image formation under gas discharge visualization (Kirlian effect) // Radio engineering and electronics. 1986. V. 31, N 4. P. 625-642

[9] Korotkov K. G. Registration of energy-informational interaction by means of gas discharge sensor // Biomed. informatics: Collection of articles. St. Petersburg, 1995. p. 197-206.
[10] Tiller W. A gas Discharge Device for Investigating Focused Human Attention. J Sci Exploration, 1990, v. 4, N 2
[11] Gudakova G. Z., Galinkin V. A., Korotkov K. G. Investigation of characteristics of gas discharge glow of microbiological cultures // J. of applied spectroscopy. 1988. V. 49, N 3. p. 412-417. Gudakova G. Z., Galinkin V. A., Korotkov K. G. Investigation of phases of growth of fungus of C. quilliermondy culture with gas discharge visualization technique (Kirlian effect) // Mycology and phytology. 1990. V.2, N 2. p. 174-179
[12] Mandel P. 1986. Energy Emission Analysis; New Application of Kirlian Photography for Holistic Medicine. -Synthesis Publishing Co., Germany.
[13] Korotkov K. Light After Life. Backbone publishing, NY. 1998. 190 p. ISBN 0-9644311-5-7.
[14] Dulnev G. N. Registration of phenomena of telekinesis //Consciousness and Physical Reality. 1998, v.3, p. 58-71
[15] Pi-Sunyer PX. Medical hazards of obesity. Ann. Int. Med. 1993; 1(19): 655-660
[16] Tutar E., Karadia S., Ziada K . M. Heart disease begins at a young age. Am. Heart Ass. Meeting. Abstract # 2760, November 9, 1999. 39 Lafferty B. Medicine, Health, and Society. Univ. of Washington (Seattle) School of Medicine Lecture Course Hubio 555. February 1999.
[17] Prusiner S. B. Prions. Scientific American, 1995; 272 (1): 48-57.
[18] Crick F. The Astonishing Hypothesis. The Scientific Search for the Soul. A Touchstone Book. NY, 1995
[19] Korotkov K., Korotkin D. Concentration dependence of gas discharge around drops of inorganic electrolytes. J of Applied Physics, V. 89. 2001. N 9, pp. 4732-4737
[20] Bell I., Lewis D.A., Brooks A.J., Lewis S.E., Schwartz G.E. Gas Discharge Visualisation Evaluation of Ultramolecular Doses of Homeopathic Medicines Under Blinded, Controlled Conditions. J Altern Complement Medicine, 2003, 9, 1: 25-37.
[21] Korotkov K. et. al. The Research of the Time Dynamics of the Gas Discharge Around Drops of Liquids. J of Applied Physics. 2004, v. 95, N 7, pp. 3334-3338.
[22] Erwin Schroedinger. Mind and Matter. Cambridge University Press, 1958, pp. 54-55
[23] Korotkov K. Aura and Consciousness - New Stage of Scientific Understanding. St.Petersburg, Russian Ministry of Culture. 1998. 270 p. ISBN 5-8334-0330-8.
[24] Korotkov K. Human Energy Field: study with GDV bioelectrography. Backbone publishing, NY. 2002. 360 p. ISBN 096443119X.
[25] Kapiza S., Kurdjumov S., Malinezky G. Synergetics and future forecasts. Moscow. 2001

[26] Hancock G. Fingerprints of the Gods. Crown Publishers, NY. 1995
[27] Le Roy E. Les origins humanines et levolution de lintelligence. Paris. 1928 50 Vernadski V.I. La biosphere. Paris: Alcan. 1929
[28] Ebeling V., Engel A., Faistel R. Physics of the processes of evolution. Synergetic approach. Moscow. 2001
[29] Bundzen P., Korotkov K., Nazarov I., Rogozkin V. Psychophysical and Genetic Determination of Quantum-Field Level of the Organism Functioning. Frontier Perspectives, 2002, 11, 2, 8-14
[30] Sheldrake R. The Presence of the Past. Fontana. Collins. London. 1989
[31] Korotkov K., Kuznetzov A. Model of interference structures in biology. In: Biomedical informatics: St. Petersburg, 1995, pp. 33-49
[32] Bundzen P., Korotkov K., Unestahl L.-E. Altered States of Consciousness: Review of Experimental Data Obtained with a Multiple Techniques Approach. J of Alternative and Complementary Medicine, 2002, 8, 2, 153-167.
[33] Korotkov K., Bundzen P., Bronnikov V., Lognikova L. Bioelectrographic Correlates of the Direct Vision Phenomenon. J of Alternative and Complementary Medicine . 2005, 11, 5, 885-893
[34] Tiller W., Dibble W., Fandel J. Some Science Adventures with Real Magic. Pavior Publishing. 2005.
[35] Ultra High Dilutions - Physiology and Physics. (Endler, Ed. Kluuiver Acad. Pub., 1994).
[36] M. Skarja, M. Berden, and J. Jerman, J.Appl.Phys., 84, 2436, (1998).
[37] K. Korotkov and D. Korotkin, J.Appl.Phys., 89, 4732, (2001).
[38] Bell I., Lewis D.A., Brooks A.J., Lewis S.E., Schwartz G.E. Gas Discharge Visualisation Evaluation of Ultramolecular Doses of Homeopathic Medicines Under Blinded, Controlled Conditions. J Altern Complement Medicine, 9, 1, 2003: 25-38

GDV BIOELECTROGRAPHY PUBLICATIONS

1. Boyers D.G. Tiller W.A. Corona Discharge Photography. J of Applied Physics, 1973, 44, 3102-3112.
2. Pehek J.O., Kyler K.J., and Faust D.L. Image modulation in Corona Discharge Photography. Science 1976, 194, 263-270.
3. Opalinski J. Kirlian-type images and the transport of thin-film materials in high-voltage corona discharge. J. of Applied Physics 1979, 50, 498-504.
4. Konikiewicz L.W., Griff L.C. Bioelectrography - A new method for detecting cancer and body physiology. Harrisburg: Leonard Associates Press, 1982. 240 p.
5. Tiller W. On the evolution of Electrodermal Diagnostic Instruments. The Journal of Advancement in Medicine 1:41-72, 1988.
6. Mandel P. Energy Emission Analysis; New Application of Kirlian Photography for Holistic Medicine. Synthesis Publishing Co., Germany. 1986.

1998

7. Kolmakow S., Hanninen O., Korotkov K., Kuhmonen P. Gas discharge visualization system applied to the study of non-living biological objects // J. Pathophysiology. 1998. 5. 55.
8. Korotkov K. Light After Life. Backbone publishing, NY. 1998. 190 p. ISBN 0-9644311-5-7.
9. Korotkov K. Aura and Consciousness - New Stage of Scientific Understanding. St.Petersburg, Russian Ministry of Culture. 1998. 270 p. ISBN 5-8334-0330-8.
10. Skarja M., Berden M., and Jerman I. Influence of ionic composition of water on the corona discharge around water drops. J. of Applied Physics. 1998, 84, 2436.

1999

11. Howell Caroline J. The therapeutic effect of tai chi in the healing process of HIV. International J of Alternative and Complementary Medicine. Nov 1999, pp. 16-20.
12. Kolmakow S., Hanninen O., Korotkov K., Bundzen P. Gas Discharge Visualisation and Spectrometry in Detection of Field Effect // Mechanism of Adaptive Behaviour : - Abstracts of Int. Sympos. St. Petersburg, 1999. 39-40.
13. Shaduri M.I., Chichinadze G.K. Application of bioenergography in Medicine. Georgian Engineering News. 1999, 2, 109-112.

2000

14. Bevk M, Kononenko I, Zrimek T. Relation between energetic diagnoses and GDV images. In Proc New Science of Consciousness Conference,. Ljublana, October 2000, pp. 54-58.
15. Bundzen P., Zagrantsev V., Korotkov K., Leisner P., Unestahl L.-E. Comprehensive Bioelectrographic Analysis of Mechanisms of the Altered State of Consciousness. Human Physiology, 2000, 26, 5, 558-566.
16. Bundzen P., Korotkov K., Massanova F., Kornysheva A. Diagnostics of Skilled Athletes PsychoPhysical Fitness by the Method of Gaz Discharge Visualisation Proceedings 5th Annual Congress of the European College of Sport Science. - Jyvaskyla, Finland, 2000. - P. 186.
17. Bundzen, P.V. and Korotokov, K.G. Health evaluation based on GDV parameters. In Proc International Scientific Congress on Bioelectrography, St. Petersburg, Russia, 2000, pp 5-7.
18. Kolmakov S. and Hanninen O. Fingertip Gas Discharge pattern in electro-magnetic field reflect Mental Activity, Finish Psysiological Society, Joensuu, 27, 2000.
19. Dobson Paul and O'Keffe Elena. Investigations into Stress and it's Management using the Gas Discharge Visualisation Technique International J of Alternative and Complementary Medicine. June 2000.
20. Sanches Fernando. Aura y Ciencia. Mandala Ediciones. Madrid. 2000.

2001

21. Kononenko I. Machine learning for medical diagnosis: history, state of the art and perspective. Artificial Intelligence in Medicine. 2001, 23, 89.
22. Korotkov K., Korotkin D. Concentration dependence of gas discharge around drops of inorganic electrolytes. J of Applied Physics, 2001, 89, 9, 4732-4737.
23. Russo M. Russo M, Choudhri AF, Whitworth G, Weinberg AD, Bickel W, and Oz MC. Quantitative analysis of reproducible changes in high-voltage electrophotography. J of Alternative and Complementary Medicine 2001, 7, 6, 617-629.
24. Rein G., Giacomoni P., Cioca G., Gubernick J., Vainshelboim A., Matravers P., Korotkov K. Characterization Of The Energetic Properties Of Gems Using The Gas Discharge Visualization Technique. Proceedings of the International Congress "Science, Information, Spirit", St. Petersburg, 2001. P.48.

2002

25. Bundzen P., Korotkov K. New computer technology for evaluating the psycho-physical fitness of athletes. Physical Education and Sport. Warszawa, 2002, 46 (1), 392-393.
26. Bundzen P., Korotkov K., Nazarov I., Rogozkin V. Psychophysical and Genetic Determination of Quantum-Field Level of the Organism Functioning. Frontier Perspectives, 2002, 11, 2, 8-14.

27. Bundzen P., Korotkov K., Unestahl L.-E. Altered States of Consciousness: Review of Experimental Data Obtained with a Multiple Techniques Approach. J of Alternative and Complementary Medicine, 2002, 8 (2), 153-167.
28. Gibson s. The effect of music and focused meditation on the human energy field as measured by the gas discharge visualisation (GDV) technique and profile of mood states. Thesis of a dissertation submitted to the faculty of HOLOS university graduate seminary. April 2002.
29. Korotkov K. Human Energy Field: study with GDV bioelectrography. Backbone publishing, NY. 2002. 360 p. ISBN 096443119X.
30. Measuring the Human Energy Field: State of the Science. Ed. R.A. Chez. National Institute of Health, Samueli Institute, Maryland, 2002.
31. Roberts N-R. Parallel investigation of the meridian stress assessment (msa-21) and the gas discharge visualization devices: can they measure the effects of acupuncture treatment on the body's energy state? Thesis of a dissertation submitted to the faculty of HOLOS University Graduate Seminary. March 2002.
32. Krizhanovsky E., Korotkov K., Borisova M., Matravers P., Vainshelboim A. Time dynamics of gas discharge around the drops of liquids// Proceedings of the International Congress "Science, Information, Spirit", St. Petersburg, 2002. P.54-56.
33. Bascom R, Buyantseva L, Zhegmin Q, Dolina M, Korotkov K: Gas discharge visualization (GDV)-bioelectrography. Description of GDV performance under workshop conditions and principles for consideration of GDV as a possible health status measure; in Francomano CA, Jonas WB, Chez RA (eds): Proceedings: Measuring the Human Energy Field. State of the Science. Corona del Mar, CA, Samueli Institute, 2002, pp 55-66.
34. Korotkov K, Donlina MY, Bascom R: Appendix: translation of Russian documents related to GDV; in Francomano CA, Jonas WB (eds): Proceedings: Measuring the Human Energy Field. State of the science. Corona del Mar, CA, Samueli Institute, 2002, pp 90-156.

2003

35. Bell I., Lewis D.A., Brooks A.J., Lewis S.E., Schwartz G.E. Gas Discharge Visualisation Evaluation of Ultramolecular Doses of Homeopathic Medicines Under Blinded, Controlled Conditions. J of Alternative and Complementary Medicine, 2003, 9, 1: 25-37
36. Korotkov K. Where Do We Go? Frontier Perspectives. 2003, 12, 3, 30-37.
37. Musiol M-J. Corps de Lumiere. Bodies of Light. Axe Neo-7 Art. 2003. ISBN 2-922794-03-2
38. Vepkhvadze R., Gagua R., Korotkov K. et. GDV in monitoring of lung cancer patient condition during surgical treatment//Georgian oncology. Tbilisi. 2003. № 1(4). p. 60.
39. Giacomoni P., Hayes M., Korotkov K., Krizhanovsky E., Matravers P., Momoh K.S., Shaath N. and Vainselboim A. Investigation of Essential Oils and Synthetic Fragrances using the Dynamic Gas Discharge Visualization Technique//World Perfume Congress. Seoul. Korea. 2003, p18.

40. Giacomoni P., Hayes M., Korotkov K., Krizhanovsky E., Matravers P., Momoh K.S., Peterson P., Shaath N., Vainselboim A. Study of cultural aspects of cosmetology using the dynamic gas discharge visualization technique// Proceedings of the International Congress "Science, Information, Spirit", St. Petersburg, 2003. P.95.
41. Vainshelboim A.L., Hayes M.T., Korotkov K., Momoh K.S. Investigation of Essential Oils and Synthetic Fragrances Using the Dynamic Gas Discharge Visualization Technique IFSCC Conference 2003; Seoul, Korea. Proceeding Book Part 1. pp. 431-43.

2004

42. Ahmetely G., Boldireva U. et. al. Allergy etiology diagnostics using Gas Discharge Visualization Technique. Proceedings of St. Petersburg Military Medical Academy. St. Petersburg, 2004.
43. Cioka G, Korotkov K, Giacomoni PU, Rein G, Korotkova A: Effects of exposure to electromagnetic fields from computer monitors on the corona discharge from skin; in Korotkov K (ed): Measuring Energy Fields State of the Science. Fair Lawn, NJ, Backbone, 2004, pp 183-192.
44. Dobson P, O'Keefe E: Research into the efficacy of the gas discharge visualisation technique as a measure of physical and mental health. http://wwwkirlianorg/gdvresearch/experiments/paulelena/refhtml 2004.
45. Gibson SS: Effect of listening to music and focused meditation on the human energy field as measured by the GDV and the profile of mood states (POMS); in Korotkov K (ed): Measuring Energy Fields· State of the Science. Fair Lawn, NJ, Backbone, 2004, pp 209-222.
46. Korotkov K. Experimental Study of Consciousness Mechanisms with the GDV Bioelectrography. In: Science of Whole Person Healing. Volume 2. Rustum Roy (Ed.). New York, Lincoln, Shanghai. 2004. pp. 152-184.
47. Korotkov K., Krizhanovsky E., Borisova M., Hayes M., Matravers P., Momoh K.S., Peterson P., Shiozawa K., and Vainshelboim A. The Research of the Time Dynamics of the Gas Discharge Around Drops of Liquids. J of Applied Physics. 2004, v. 95, N 7, pp. 3334-3338.
48. Korotkov K., Williams B., Wisneski L. Biophysical Energy Transfer Mechanisms in Living Systems: The Basis of Life Processes. J of Alternative and Complementary Medicine, 2004, 10, 1, 49-57.
49. Measuring Energy Fields: State of the Art. GDV Bioelectrography series. Vol. I. Korotkov K. (Ed.). Backbone Publishing Co. Fair Lawn, USA, 2004. 270 p. ISBN 097420191X
50. Owens J, Van De Castle R: Gas discharge visualization (GDV) technique; in Korotkov K (ed): Measuring Energy Fields State of the Science. Fair Lawn, NJ, Backbone, 2004, pp 11-22.
51. Rubik B: Scientific analysis of the human aura. In Korotkov K (ed): Measuring Energy Fields State of the Science. Fair Lawn, NJ, Backbone, 2004, pp 157-170.
52. Senkin V.V., Ushakov I.B., Bubeev U.A. Bioelectrographic criteria of overload

tolerance of summer pilot team in centrifuge expert test // Proceedings of the conference "Neurobiotelekom" SPb, 2004, pp 69-70
53. Vainshelboim A.L., Hayes M.T., Korotkov K., Krizhanovsky E., Momoh K.S. Investigation of natural and synthetic flavors and fragrances using the dynamic gas discharge visualization technique. Proceedings of PITTCON Conference. Chicago 2004. p. 14900-900.
54. Vainshelboim A.L., Hayes M.T., Korotkov K., Momoh K.S. Intrinsic Energy of Odorant and Olfactory Responses Using Gas Discharge Visualization International Congress of Systematic Medicine. Caracas, Venezuela. January 21-22, 2005. pp. 236-238.
55. Vainshelboim A.L., Hayes M.T., Korotkov K., Momoh K.S. Investigation of Natural and Synthetic Flavors and Fragrances Using the Dynamic Gas Discharge Visualization Technique. PITTCON Abstract 2004. CD-ROM. 14900-900.
56. Vainshelboim A.L., Hayes M.T., Korotkov K., Momoh K.S. Observing the Behavioral Response of Human Hair to a Specific External Stimulus Using Dynamic Gas Discharge Visualization TRI/Princeton Conference on Applied Hair Science. Princeton, New Jersey. June 9-10, 2004. Book of Abstracts.
57. Vainshelboim A.L., Hayes M.T., Korotkov K., Momoh K.S. Observing the Behavioral Response of Human Hair to a Specific External Stimulus Using Dynamic Gas Discharge Visualization Journal of Cosmetic Science. Proceedings of the First International Conference on Applied Hair Science. Full Manuscript. Princeton, New Jersey. June 9-10, 2004. pp. S91-S104.
58. Vainshelboim A.L., Hayes M.T., Korotkov K., Momoh K.S. Observing the Behavioral Response of Human Hair to a Specific External Stimulus using Dynamic Gas Discharge Visualization IFSCC 3rd Congress. Orlando, FL 2004. Abstract.
59. Vainshelboim A.L., Hayes M.T., Momoh K.S. Aveda advertisement - Tourmaline Charged Radiance Fluid Jane Magazine. August 2004. pp. 24-25
60. Vainshelboim A.L., Hayes M.T., Momoh K.S. Aveda GDV Research Measures Raw Material Energies "The Rose Sheet" Toiletries, Fragrances, and Skin Care. Vol. 25, No. 16. April 19, 2004. pp. 4
61. Vainshelboim A.L., Hayes M.T., Momoh K.S. Investigation of Energetical Properties of Holistic Cosmetic Materials and Products PCITX Personal Care Ingredients & Technology Exposition. April 14, 2004.
62. Vainshelboim A.L., Hayes M.T., Momoh K.S. New Approaches to Testing Natural Fragrances and Flavors. Happi Magazine. January 2005.

2005

63. Bundzen P. V., Korotkov K. G., Korotkova A. K., Mukhin V. A., and Priyatkin N. S. Psychophysiological Correlates of Athletic Success in Athletes Training for the Olympics Human Physiology, Vol. 31, No. 3, 2005, pp. 316-323. Translated from Fiziologiya Cheloveka, Vol. 31, No. 3, 2005, pp. 84-92.
64. Hacker G.W., Pawlaka E., Pauser G., Tichy G., Jell H., Posch G., Kraibacher G., Aigner A., Hutter J. Biomedical Evidence of Influence of Geopathic Zones

on the Human Body: Scientifically Traceable Effects and Ways of Harmonization. Forsch Komplementarmed Klass Naturheilkd. 2005.
65. Korotkov K. Champs D'Energie Humaine. Resurgence Collection. Belgique. 2005.
66. Vainshelboim A.L., Hayes M.T., Korotkov K., Momoh K.S. Electric and Magnetic Field and Electron Channeling in Human Hair. IFSCC 23rd Congress. Florence, Italy September 2005.
67. Vainshelboim A.L., Hayes M.T., Korotkov K., Momoh K.S. GDV Technology Applications for Cosmetic Sciences IEEE 18th Symposium on Computer-Based Medical Systems (CBMS 2005). Dublin, Ireland. June 2005.
68. Vainshelboim A.L., Hayes M.T., Korotkov K., Momoh K.S. Investigation of Conscious and Subconscious Reactions to Essential Oil Blends ISOEN Olfaction and Electronic Nose 11th International Symposium. Barcelona, Spain. April 13-15, 2005. Poster presentation.
69. Vainshelboim A.L., Hayes M.T., Korotkov K., Momoh K.S. The New Investigation of Specific Aqueous Systems Using Dynamic GDV-Graphy PITTCON. Orlando, FL. March 3, 2005. Abstract presentation.
70. Vainshelboim A.L., Hayes M.T., Korotkov K., Momoh K.S. Utilization of Powdered Gemstones in Oil-Based Formulations. IFSCC 23rd Congress. Florence, Italy September 2005.
71. Korotkov K., Bundzen P., Bronnikov V., Lognikova L. Bioelectrographic Correlates of the Direct Vision Phenomenon THE JOURNAL OF ALTERNATIVE AND COMPLEMENTARY MEDICINE . Volume 11, Number 5, 2005, pp. 885-893
72. Vainshelboim A.L., Hayes M.T., Momoh K.S. Bioelectrographic Testing of Mineral Samples: A Comparison of Techniques. Journal of Alternative and Complementary Medicine. 2005: Vol. 11, No. 2, pp. 299-304.

2006

73. Korotkov. K. Geheimnisse des lebendigen Leuchtens. Herstellung Leipzig, Germany, 2006, pp. 142
74. Korotkov. K., Carlos Mejia Osorio. La Bioelectrografia. Una vision a la medicina del siglo XXI. Baranquilla, Colombia. 2006. pp. 67

Proceedings of the International Congresses "Science, Information, Spirit". St. Petersburg. 1998 - 2006.
For the latest news and information, please, visit: www.korotkov.org

NOTES

NOTES

SPIRAL TRAVERSE
Journey into the Unknown

Konstantin Korotkov

Computer page-proof
OOO "D`ART"
d_art@savog.spb.ru

Design and cover
Alisa Pakhomova

Backbone Publishing Company
P.O. Box 562, Fair Lawn, NJ 07410, USA
FAX: 201 670 7892
bbpub@optonline.net
www.backbonepublishing.com

Printed in Russia
Size 60x90 1/16. Offset printing. Offset paper.
Volume 18 printer sheets. Edition: 1000. Order № 109.

Printed from ready slides in OOO "SAVOG"
Revolyutsii sh. 69, lit. A, St. Petersburg, 195279, RUSSIA
info@savog.spb.ru